Silvicultures - Management and Conservation

Edited by Fernando Allende Álvarez,
Gillian Gomez-Mediavilla
and Nieves López-Estébanez

Published in London, United Kingdom

IntechOpen

Supporting open minds since 2005

Silvicultures - Management and Conservation
http://dx.doi.org/10.5772/intechopen.83683
Edited by Fernando Allende Álvarez, Gillian Gomez-Mediavilla and Nieves López-Estébanez

Contributors
Gillian Gomez-Mediavilla, Fernando Allende Álvarez, Nieves López-Estébanez, Helen Read, Vikki Bengtsson, Valentina Pescini, José Muñoz-Rojas, Teresa Pinto-Correia, Egon Noe, Martin Hvarregaard Thorsøe

First published in London, United Kingdom, 2019 by IntechOpen
IntechOpen is the global imprint of INTECHOPEN LIMITED, registered in England and Wales, registration number: 11086078, The Shard, 25th floor, 32 London Bridge Street
London, SE19SG – United Kingdom
Printed in Croatia

British Library Cataloguing-in-Publication Data
A catalogue record for this book is available from the British Library

Additional hard and PDF copies can be obtained from orders@intechopen.com

Silvicultures - Management and Conservation
Edited by Fernando Allende Álvarez, Gillian Gomez-Mediavilla and Nieves López-Estébanez
p. cm.
Print ISBN 978-1-83880-719-1
Online ISBN 978-1-83880-720-7
eBook (PDF) ISBN 978-1-83880-721-4

For EU product safety concerns:
IN TECH d.o.o., Prolaz Marije Krucifikse Kozulić 3, 51000 Rijeka, Croatia, info@intechopen.com or visit our website at intechopen.com.

Meet the editors

Dr. Fernando Allende Álvarez is an associate professor of Regional Geographical Analysis at Universidad Autónoma de Madrid (UAM) and has a PhD in Geography (2007). He combines modelling and field studies of vegetation and climate using quantitative and qualitative data. His research focuses on woodland ecology, mountain climates, geomorphology, and applied techniques in remote sensing and geographical information systems (GIS). He has collaborated in multidisciplinary research groups with different universities and institutions in Europe, Latin America, and the USA.

Gillian Gomez-Mediavilla is a researcher at the Geography Department of UAM. She is an Environmental Science graduate and has research experience in the history and management of forest and rural areas, agricultural and cultural landscapes, changes in land use, vegetation dynamics, as well as techniques related to GIS and environmental cartography. Her PhD thesis addresses the forest landscape of Fraxinus angustifolia pollards in Spain and its relevance in terms of biodiversity, cultural heritage, and economic and social values.

Dr. Nieves López-Estébanez, PhD in Geography (2003), works as an associate professor in regional geographical analysis at UAM. She is a specialist in Mediterranean vegetation with special interest in mountain ecosystems and cultural use of woodlands. She has been an invited researcher at the University of Costa Rica where she investigated agro-landscapes in the Volcanic Central Range. She is also an active member of multidisciplinary research groups on landscapes.

Contents

Preface

Oliver Rackam in his work *Woodlands* (Rackam, 2006) considers:

> *(...) for a thousand years woodlands were relatively stable. Woods were assaulted by outside forces, grubbed out into farmland or grazed away into heath. There were gradual changes from within: encoppicement of wood pastures, lengthening of coppice cycles, shifts in the balance between underwood and timber, or (in the Scottish Highlands) increased grazing and the change from cattle to sheep (...).*

This idea is the central axis of this monograph dedicated to the pollarded forests in Europe. Those who have maintained an equilibrated silvosystem and rational uses. The work that we present is a wonderful trip to understand some of the more ancient forest cultures in Europe. The journey will take us from the English forests of *Quercus* ssp. to the mixed forests of *Abies* ssp., *Fagus* ssp., and *Quercus* ssp. in the Apennines. We will reach the centre of the Iberian Peninsula with the forest of *Fraxinus angustifolia* in the Spanish Central System. We will finish with the Portuguese *Montados* of *Quercus rotundifolia* and *Quercus suber*.

We start in England with Helen Read and Vikki Bengtsson, who masterfully talk about the pasture-tree binomial or wood pastures and their management in the South East of England. They include an interesting idea in their conclusions:

> *New uses for the trees in combination with grazing, to make wood pastures both attractive and sustainable, are perhaps the way forward if we are to ensure that there are wood pastures with old trees in the future.*

In Italy, between Liguria, Tuscany, and Emilia-Romagna, in the middle of the Apennines, Diego Moreno, Roberta Cevasco, Valentina Pescini, and Nicola Gabellieri reconstruct the influence of man on woodlands through archaeology. The authors offer an enriching view from a historic perspective, field knowledge, and contributions of palynology and anthracology. They establish a nexus between the documented and deduced past through indirect data. This lets them highlight that:

> *The medieval and post-medieval history of the wooded resources of the Mediterranean mountains appears dominated by animal production; a complex agro-silvo-pastoral system concealed by the development of nineteenth and twentieth-century forestry sciences and allied forestry law regime that brought the end to the customary regime, which was used to regulate woodland and commons' resources access.*

In the centre of the Iberian Peninsula, we find the Mediterranean continental mountain, where the most productive forest is located in the wet bottom valleys. Fernando Allende, Gillian Gomez-Mediavilla, and Nieves López-Estébanez approach *Fraxinus angustifolia* formations from a threefold perspective that combines history, importance of the physical environment, and property.

Finally, José Muñoz-Rojas, Teresa Pinto-Correia, Martin Thorsoe, and Egon Noe take us to the Alentejo agro-silvo-pastoral systems based on the use of *Quercus suber* and *Quercus ilex* subsp. *rotundifolia*. For them, the *montado* is:

> *A system that is characterized by its great potential for improving sustainability standards by helping reconcile farming productivity with biodiversity conservation, landscape-cultural heritage protection and local economic development.*

After reading all these works, the reader will understand the need to strengthen the research and knowledge of pruned and pollarded forests, to outline better management and establish informative, awareness, and value enhancement actions for ancient European forests.

Fernando Allende Álvarez, Gillian Gomez-Mediavilla, and Nieves López-Estébanez,
Geography Department, Universidad Autónoma de Madrid, Cantoblanco, Spain

Chapter 1

The Management of Trees in the Wood Pasture Systems of South East England

Helen Read and Vikki Bengtsson

Abstract

This chapter outlines the history and past management of trees within the wood pasture systems of South East England. Changes over time are discussed, and the challenges that the trees now face are outlined along with some potential solutions. Wood pasture was a common and traditional form of management in South East England although the conservation significance of it has only recently been realised. The types of wood pasture included wooded commons, Forests and parks, all of which have quite precise historical meanings. Many trees in wood pastures were managed as pollards, probably mainly for fuel wood, but some were open-grown. The number of trees has declined, and the area of wood pasture has diminished due to development pressure and agricultural intensification. Despite this, the area remains important in a European context for the number of old trees. In addition, lack of traditional management is a threat to tree and wood pasture survival. Restoration of grazing using traditional livestock is an important first step. New skills are required to work on trees that have been left many years out of a regular pollarding cycle, and new uses for the products will be important to help these trees become relevant again.

Keywords: wood pasture, pollards, England

1. Introduction

In conservation terms wood pasture is only a relatively recently recognised habitat in the UK, in contrast to countries like Spain where the *dehesa* is well known and still actively managed. Following the 1992 Rio Convention on Biodiversity, the UK focused on conservation action through the development of biodiversity action plans (BAPs) for species and habitats. Wood pasture was included in the form of 'wood pasture and parkland'. One of the first problems for the steering group setup to oversee the BAP was defining the habitat.

At the time that the UK joined the EU, wood pastures were not listed as a CORINE habitat type, so that Special Areas of Conservation under the Habitats and Species Directive could only be granted to wood pasture sites in the UK under other, usually woodland, habitats. Later, at the time of the accession of several Scandinavian countries, like Sweden, to the EU, two extra habitats were proposed and accepted, Fennoscandian wooded meadows (6530*) and Fennoscandian wooded pastures (9070). Some of these are very similar to areas in the UK, but

IntechOpen

designation under these categories was not permitted to countries that accessed before these countries. This remains the case today (see [1] for more details and a list of the wood pasture sites designated as Special Areas of Conservation SAC).

Despite the current situation with SAC sites, the English agri-environment schemes fund both the maintenance and restoration of wood pastures. The former BAP steering group has morphed into a technical advisory group and has produced guidance in the form of videos [2] including one for advisers, to help them recognise and value wood pastures and increase their knowledge of the appropriate management. Other management films and training materials focusing on the management of the habitats and, in particular the trees, have been developed through the Ancient Tree Forum [3, 4] and an EU project [5].

The irony is that wood pasture is probably one of the oldest forms of land management in Southern England and it was probably extremely widespread and common in the past. It is also most likely that wood pasture mimicked the appearance of at least parts of the English landscape in times before humans altered it so substantially [6].

Wood pasture and parkland are mosaic habitats valued for their trees, especially veteran and ancient trees, and the plants and animals that they support. Grazing animals (either domestic livestock or deer) are fundamental to the existence of this habitat along with flowers providing nectar sources and open grassland or heathland ground vegetation [7]. In this chapter we outline the different origins of the wood pasture systems in South East England and look at the historical management of the trees. Some of the changes are then discussed as well as the problems that the trees now face. Finally, we look to the future and suggest some potential solutions. Where possible we have given details of some case studies which present some key sites and trees in a little more detail.

2. Historical background

Trees started colonising the UK following the end of the last ice age in around 8200 BC. The density of the tree cover that developed has recently become a matter of conjecture, with the traditional view that this was dense and continuous [8] being challenged in recent years due to the work in other parts of Europe (for example, see [6]). Kirby and Watkins [9] consider that, prior to the Neolithic period, tree cover in England probably ranged between 60 and 90% but with some parts being much more open. Although undoubtedly early man had some impact on local areas, around 3100 BC more extensive evidence of human impact on the English landscape starts to be found. This includes more widespread tree felling, and finds of ancient trackways from Neolithic and Bronze Ages indicate that wood was being used and quite probably was grown for use [10]. Between the birth of Christ and 1200 AD, the English countryside changed into a 'modern' landscape with villages, fenced areas and orderly management of woodland. During this period the Romans and then Anglo-Saxons had large impacts on the land, including periods when the tree cover increased [11]. They were followed by the Normans who also had a major influence on the land. In 1086, just 20 years after the Norman Conquest, the first inventory was taken from the English landscape on a huge scale which included descriptions of the size and location of woodland, as well as other features, for each parish (the smallest administrative unit in England). These are still accessible today in the *Domesday Book*. Unfortunately, it is not always easy to interpret every entry in the *Domesday Book*, and there are regional variations in the way different land management is described; however, it is evident that both coppiced woods (*silva minuta*) and pasture wood (*silva pastilis*), which has been described [12] as being 'park-like with plenty of grass', existed. Probably, *silva pastilis* constituted the greater part of woodland at this time [11]. By the thirteenth century, it is considered [12] that these two types of woodland were clearly differentiated and it was relatively rare for woods to be converted from one to the other (although this clearly did happen in places [13]), and most woods were now managed as coppice [14].

The management of woodlands in the medieval landscape was considered to be intensive and relatively stable, with replacement trees grown in place of any cut down. Woods were enclosed by banks and ditches combined with a fence or hedge. Wood pastures were rather less stable than the woodlands, and many had been lost by the thirteenth century.

In other parts of the UK, particularly in upland areas, the land use history and patterns were a little different, and the trees and wood pastures were subject to different pressures [1, 10, 14] which is why in this chapter, we confine our discussion to South East England, especially the area around London, which still has some particularly notable places for trees and wood pasture to this day.

2.1 Types of wood pasture and where they were found

There are many different types of wood pasture in South East England, but the most frequent were those associated with the historic land uses of wooded commons, Royal Forests and parks. In addition, fields with trees can be considered a form of wood pasture. These different types are outlined below and some case studies of typical examples.

2.1.1 Wooded commons

Common land was abundant in the English countryside in the twelfth and thirteenth centuries, and probably every parish had some, even today there is over 1.3 million acres [15] (526,000 hectares) of common land remaining in England and Wales. Commons were owned by a person, often the Lord of the Manor or the owner of a large estate. Although it was usually a wealthy person who owned the common, villagers could use the land, subject to certain rules and regulations. Typically, commons provided fuel, small-scale wood and grazing for the poor of the village [16] with cattle, sheep, horses, geese and pigs, all turned out on them. Thus, they were areas where there was a high intensity of management, and they were crucial to the survival of many people.

Case study: Ashtead Common. At 200 ha in size and on heavy clay soils, Ashtead Common has over 2300 *Quercus* sp. pollards [17] (of which about half are still alive) (**Figure 1**). As part of the Manor of Ashtead, it was probably used for providing wood and grazing land for the local people. Earthworks in the wood indicate Iron Age working, and there are also a Roman villa and tile works. Pollarding probably ceased around 150 years ago, and recently a work programme to reduce the crowns of the trees has started to try to stop them falling apart from the heavy weight of the branches. Ashtead Common is owned and managed by the City of London Corporation.

2.1.2 Parks

The word park has conjured up images of very different landscapes in England over the years. In Anglo-Saxon times, it was an area of land with a fence around it. From 1066 until the eighteenth-century parks were places where wild animals like deer (of various species) or semi-wild animals such as feral cattle were kept for the main purpose of hunting. These proliferated from the thirteenth century, and it has been calculated that in lowland England there may have been roughly one park to every four parishes or every 38 square kilometres [12]. The animals were kept

Figure 1.
Lapsed Quercus *sp. pollards at Ashtead common.*

in the parks by a deer-proof boundary, the most characteristic being a 'park pale' or palisade fence of cleft oak, very expensive to build and maintain. The park pale could also be a wall, but this was less likely in South East England. Parks varied in size and, because of their cost, were the preserve of rich people and those that could afford such a status symbol. Inside the park pale, the land was grazed, but it could range between dense woodland and completely open areas. Typically, much of the area was grassland with scattered trees, sometimes pollarded, but this was not universal.

In the seventeenth and eighteenth centuries a new type of park evolved, that of a formally designed landscape park. Famous designers included Lancelot (Capability) Brown and Humphry Repton, amongst others, who were employed by wealthy landowners to make their estates attractive. In many cases they imposed these on an older park, frequently retaining features such as the older trees [12]. Most of the landscaping involved physically moving soil to create lakes and mounds and the planting of trees in specific patterns such as clumps or avenues [12]. Parts of these parks continued to be grazed by livestock or deer.

Case study: Windsor Great Park was the largest deer park in England in medieval times and today contains just under 2000 ha [18]. The modern boundary of the park follows the old park pale in many places although today it is part of a large area adjacent to Windsor Castle, much of which is open to the public and also comprises areas of working farmland and forestry. More than 7000 significant trees have been identified across the whole of the estate [19], but the *Quercus* spp. are the most abundant and impressive with many exceeding 6 m in girth (**Figure 2**). The land under the trees is mostly acid grassland and is grazed by deer. Several avenues form part of the overall complex which also includes ponds and streams. The park is partly owned by the state and managed as part of the Crown Estate.

2.1.3 Royal Forests

The word Forest has a very specific meaning in England, as an area of land where specific laws operated [12]. Thus, it was not usually enclosed, like a park,

Figure 2.
Old Quercus *sp. tree at Windsor Great Park.*

but included a wide range of different land uses and sometimes even towns. Later in time the term Forest was used more to describe that part of the land that was not woodland, farmland or built upon [12]. This could be open grassland or heathland, but much was also probably wood pasture. A Royal Forest was an area where the king had rights, including the right to hunt. There was often a great deal of administration associated with the Forest, and many people were employed to supervise the interest of the Crown (royalty) in the land. These administrators gained rights through their positions, and these included rights to pasture animals, take wood and timber, etc. The land was not necessarily owned by the Crown, but the owners were prohibited from doing certain things with their own land, and this frequently included cutting down their own trees [12]. Chases are equivalent to Forests but were privately owned not part of the Crown land. At one time Forests and chases covered approximately one fifth of England [12].

Case study: Hatfield Forest is considered to be the most intact example of a medieval hunting Forest remaining in the UK today. It was a compartmentalised Forest, meaning that there were different areas within the Forest that were clearly marked out and the tree management was rigidly defined. About 220 hectares of the total 400 hectares was coppice cut on an 18-year cycle. The cut areas were fenced to keep all browsing animals out for the first six years, after which deer were admitted, and then after further three years all animals were admitted. The remaining area of the Forest was open 'plain' of grassland with pollarded trees and scrub (see [20] for more details). Today, some 800 pollards remain (**Figure 3**), including more tree species as pollards than any other site in England and *Carpinus betulus* as the most frequent. Today, Hatfield Forest is owned and managed by the National Trust.

2.1.4 Fields with infield trees

Field systems present in England today do not have a single origin. The oldest systems (sometimes called ancient countryside) are Bronze Age or earlier and difficult to date. Others were created post the Norman conquest of 1066 and were based on a mosaic of strips or parcels of land within a parish's, or manor's, common-field system. These initially had few or no hedges. Another field origin was the "inclosure" of the commoners' grazing lands called 'commons', mostly rough pasture but also some

Figure 3.
Carpinus betulus *pollards at Hatfield Forest.*

woodland. This affected over 20% of all parishes in England [12]. Some enclosure was piecemeal and illegal, but Acts of Parliament made the process legal from 1700 onwards. Whenever and wherever strip fields or common grazing land was enclosed, the land was divided up into fields, bounded by hedges, in an *ad hoc* manner in old field systems, more rigidly planned and planted under legal inclosure.

Today, relatively few English fields contain trees, but old maps and documents indicate that this was not always the case and most fields had trees in them, more in pasture than arable. In the Ancient Countryside "clayland" region of Suffolk, Norfolk and part of Essex and elsewhere in South East England, trees were managed as pollards or shreds in several systems; scattered infield trees line 3–4 trees deep, around the margins of fields, forming a type of linear wood pasture system [13], and trees within the hedge line around the fields, sometimes in very high numbers (up to a pollard every 10/15 m), a pattern throughout much of this region, up to about the mid-nineteenth century. Old maps and farm sale documents indicate that almost all fields historically had trees within them, but these mostly disappeared as tractors arrived throughout the agricultural intensification of the twentieth century.

Case study: Pollarded trees in the hedges of Goswold Hall, Suffolk. Goswold Hall is a 150 ha arable farm with mixed species hedges around the small Bronze Age fields (a nearby Roman road cuts diagonally across them) and several green lanes or ancient trackways pass through the farm. Prior to the 1970s, there were over 100 *Ulmus* sp. pollards within the hedges, but these all died from Dutch elm disease. *Quercus* sp. and *Fraxinus excelsior* pollards were also abundant, and many of these still survive; most pollarding had lapsed by the mid-nineteenth century (due to the availability of coal for fuel brought by canals or railway), but a few were cut until the late twentieth century (**Figure 4**). Pollards were cut on a 12–15-year cycle (still mostly for firewood); the farm workers were given the wood from a tree and could sell the wood if they did not need it themselves, but they had to cut the trees in their own time [3].

2.1.5 Summary

The density of trees across the historic landscape of Southern England must have varied widely, with some fields containing many infield trees, while other areas of commons, Forests or parks included everything from dense woodland to areas with

Figure 4.
Pollarded Quercus *sp. trees in an ancient hedge at Goswold hall.*

almost no trees. Thus, 'wood pasture' has not acquired a separate identity in the UK but was characteristic of several different forms of land management. It was also potentially difficult to delineate because it blended seamlessly into open habitats like pasture or places with dense tree cover like woodlands within the same park or Forest. Even on a common, there could be very open areas with almost no trees and contrasting places where the density of trees or pollards was so great that they were in effect like woodlands.

There was probably also a high degree of convergence between the different landscapes so that some places with hedges and infield trees could be potentially quite similar to areas within commons, Forests and parks.

2.2 Management of trees in wood pasture systems

2.2.1 How and why were the trees managed?

The management, and therefore the appearance of the tree, varied within the different wood pasture systems depending on the needs of the people owning and managing the land around the tree. Below, the main traditional tree management methods are outlined, those of pollarding, coppicing and shredding, and are contrasted with open-grown trees.

2.2.2 Pollards

On common land in peaty areas, fuel could be cut from the ground as turves or peat, but in many places, it was the trees that provided it as wood [12]. Wood could be collected from the ground (sometimes leaf litter was collected too) or cut from the trees. Tree felling or coppicing was not generally permitted, but the cutting of branches from the trees as pollarding often was. It seems likely that in most cases the cutting of pollards was not as regular or organised as coppicing, but as records describing the management practice in detail are scarce, it is difficult to know.

Wood cut from the trees was probably largely used as domestic fuel, but other uses dependent on the area might have been charcoal, small-scale building materials, fence posts, etc., and sometimes the wood was sold rather than used by the villagers themselves, for example, at Epping Forest where bundles of wood cut from *Carpinus betulus* pollards were sold for use in London's bread ovens.

The cutting of trees for leaves to feed to livestock as fodder or leaf hay is poorly documented in the UK. Hooke [11] indicates that *Quercus* spp., *Fraxinus excelsior* and *Ulmus* spp. were lopped (cut) in summer for winter fodder, and Barnes and Williamson [13] infer that leaves may well have been used for this purpose, but, while it was a known practice in upland areas such as the Lake District [21], its prevalence in South East England is unknown. Guidance from the sixteenth-century sources such as Tusser [22] and Fitzherbert (quoted in [25]) supports the traditional view that tree and hedge cutting was done in the winter. Fitzherbert specifies that pollarding was not carried out in 'sappe time', and Tusser [22] lists work according to the calendar stating that 'lopping of pollengers' [pollards] is a January task. However, it seems highly likely that, especially in dry years, branches would have been cut from the trees so that livestock could eat the leaves. While most villages probably had hay meadows which may have been grazed by livestock after the hay was cut, the grazing pressure, particularly on those areas with poor or thin soils, must have meant that leaf hay was needed, at least in some years.

The trees on common land were important also for their fruit. Some was no doubt consumed by humans, but the autumn pannage season, when the pigs were turned out into the woods, was an important period in the farming year, and some areas were referred to as swine pastures [12]. The acorns, beech nuts and other fruits and berries were crucial to the fattening up of the pigs and survival of poor people during the winter months although with *Quercus* sp. and *Fagus sylvatica* trees producing intermittently fruit, mast was not necessarily abundant every year.

Outside areas of common land, trees in hedges were also pollarded; for example, in parts of East Anglia, it is thought that at one time 60% of the *Quercus* sp. trees were pollards [13]. In some places the density must have been quite high, and it has been speculated that, because there were also coppice trees to provide fuel wood, the pollards must have either been cut on a longer rotation to provide larger-diameter wood for fencing and building or possibly on a shorter rotation to provide leaves [13].

In some Forests, trees were also pollarded because, although the King or another wealthy person had the rights to hunt deer, these places remained extremely important for the livelihoods of the local people. Hence in, for example, Hatfield and Epping Forests, large numbers of trees were pollarded [12, 23]. Epping Forest is so large that it extended into several parishes, and the Forest was divided up so that the people from a particular parish had specific areas where they could cut their wood.

Working trees (see [24]), i.e. those that were pollarded or otherwise cut to produce a product, were an essential component of the landscape for the majority of the people living in South East England. They were perhaps the typical tree of the poor man.

Case study: Historical pollarding of beech in Burnham Beeches [25]. Burnham Beeches is a wooded common that today has just under 400 ancient pollarded trees, mostly *Fagus sylvatica*, but around 20% *Quercus* spp. (**Figure 5**). In the 1930s there were 1795 pollards on 81 hectares with densities up to 88 per hectare, but the number has declined, and some 300 trees were estimated to have been lost in the last 50 years. Tree rings were examined in the 1930s, and it was determined that, at that time, the trees were 270 to 360 years old. They were first cut at the age of 25–35 years old and subsequently cut on an irregular cycle of 11–12 years, extending up to 15 in some older trees. Cutting began to decline in the eighteenth century but continued until about 1820 with the last trees cut being those that were smaller and more easily accessible. Today, Burnham Beeches is owned and managed by the City of London Corporation.

Figure 5.
Lapsed Fagus sylvatica *pollards at Burnham Beeches.*

2.2.3 Open-grown trees

Within hunting Forests, in deer parks or in designed landscapes associated with large country houses, the trees often fulfilled a different function. There was less need for them to be productive because fuel wood could be produced from specifically designated areas such as coppices (see below). So here the trees were more important as a part of the landscape. They either provided cover to ensure an exciting hunt for deer or they could simply look attractive. Many of the designed landscapes incorporated older trees, but many more planted trees in specific blocks, avenues or locations to frame views or to draw the eye to specific features. Many of these trees developed in very open and light conditions (**Figure 6**). In contrast to the pollards with their repeatedly cut, short stubby branches, these 'open-grown' trees spread their branches widely, also in contrast to trees grown in dense woodland [24]. The trees developed large, full crowns and branches spreading out horizontally which sometimes came right down to touch the ground where grazing by livestock or deer was at low density or absent.

2.2.4 Coppice

Perhaps the most well-known woodland management in Southern England, and that which there has been much written about, is coppice with standards. These were areas of woodland managed specifically for wood production in a variety of forms. Trees such as *Corylus avellana*, *Fraxinus* spp. and *Ulmus* spp. were coppiced, i.e. cut close to the ground on a regular cycle (often 8–10 years), and in many places were formalised and regulated. Wood from coppice was used for the walls of houses (as wattle and daub), hurdles, small-scale fencing, thatching spars (for holding the thatch onto roofs) and bean/pea poles, and the bark could be used for tanning leather. Often coppice plots contained 'standards' which were trees managed as maidens (i.e. not cut as coppice) and which were harvested on a longer cycle of 80–100 years for their timber (for planks) (**Figure 7**). These woods were securely enclosed by banks, ditches or fences to keep livestock and deer out

Figure 6.
Open-grown trees in a formal landscape at Burghley Park.

Figure 7.
Tilia and Corylus coppice *in a wood with standards in a Lincolnshire wood.*

of them and allow the coppice to grow, but sometimes they were let in towards the end of the cutting cycle, and at Hatfield Forest this was part of the organisation of the whole Forest [20].

2.2.5 Shredding

It is relatively unknown to what extent shredded trees were of importance in South East England although it is mentioned in documents up to 1600 [10]. Repeated cutting of the side branches up the trunk of the tree was a technique used widely in other countries such as Spain, Romania and Nepal, and it is said that this technique generates more leaves for use as fodder than pollarding. Whereas pollards can still be quite easily identified some 200 years after they were last actively managed, this is not so true of shredded trees. There is documented evidence in the form of photographs that these were certainly present [4, 13], but their extent and frequency are unknown.

3. Changes between then and now

3.1 Losses of trees/decline in management

England is now one of the least wooded countries in Europe. In the late nineteenth century, there was just 4% woodland although this has now increased to 13% tree cover [14] and continues to rise due to large-scale tree planting by organisations like the Woodland Trust. Trees outside woodlands, those in farmland and wood pastures, were not immune from this decline although in many cases it is less easy to document. Grubbing up of hedges was extremely widespread with the aim of increasing the size of fields for modern and large agricultural machinery, and it is thought that just 1–2% of the hedges once in the UK are still remaining [26–28]. Hedgerow loss also resulted in the loss of the trees within them. In one county the loss of trees between the 1880s (as recorded on a series of maps that showed individual trees) and aerial photographs taken in 1946 varies between 25 and 70% with the average loss being 50%. Parishes may have had up to 400 trees per square kilometre with a girth over 2′ (60 cm) in their hedges or fields (these figures do not include the woodland trees) [13]. From the late nineteenth century onwards, large-scale felling of pollards was probably quite common as they were viewed as 'unnatural' and perhaps seen as relics of a backward peasant farming system [13]. In Epping Forest the abundant but small *Carpinus betulus* pollards (**Figure 8**) in particular were considered ugly, and for this reason, they were very heavily thinned, an operation that involved the felling of tens of thousands of pollards over most of the Forest [23, 29]. The density of trees is still very high today, so it is incredible to think what it must have been like in the past.

Reasons for the extensive losses include agricultural intensification, urbanisation and conversion to commercial forestry but also abandonment. Once the tree products were no longer valued, they stopped becoming relevant to the local people.

Figure 8.
Example of a group of small pollarded trees at Epping Forest, where pollarding may have been on an industrial scale.

This may not have resulted in them being felled but in the long term led to their demise through shading out by younger trees or falling apart once regular pollarding ceased. In addition, human values have changed over time, and today we appreciate lapsed pollards as indicators of a previous culture and farming system even if we do not necessarily consider them beautiful.

3.2 Why do we have trees left today?

Despite the massive decline in trees and woodland in England, it is surprisingly one of the best places in Europe to see ancient and other veteran trees. Farjon [29] considers that the only other place in Europe with substantial numbers of *Quercus robur* or *Quercus petraea* 6 m or more in diameter at breast height is Southern Sweden (although he may have underestimated those in Spain) and that for *Quercus* spp. over 9m there are more in England than in the rest of Europe combined. Being an island, Britain tends to have an impoverished fauna and flora in comparison with other mainland European countries at the same latitude, but the saproxylic fauna can be comparatively rich [30]. Farjon [29] has noted that there seems to be no consistent environmental factors that could be responsible, so why are there so many old trees remaining today? The reasons for this and for the current distribution of veteran trees in the UK have been speculated and may include the following.

3.2.1 Places with poor soils

Commons were usually established on the poorer quality land in any parish. The richer soils were cultivated, but the areas with thin or wet or heavy clay-dominated soils were less easy to till and therefore became the rough grazing for the community. Trees grew on many of these areas and were managed for example, by pollarding, but during periods of agricultural intensification, they were less attractive for conversion. There are examples of open commons being ploughed during the World War II in order to boost food production but abandoned after 1 or 2 years because they were so hard to work and unproductive.

3.2.2 Stable (rich) landownership

The landownership structure in the UK and the inheritance system, where the eldest (usually son) inherits the entire estate, tended to keep large estates intact. Deer parks and chases therefore were less likely to be broken up, and commons also frequently remained in the same family for generations. Fashions may have changed over the years, and former 'wild' deer parks became more formally landscaped, but individual trees within them were often retained. The landscaping movement also planted trees in open conditions, either singly or in small groups, which boosted the number of open-grown trees for the future.

3.2.3 Green lungs for London

The farsightedness seen by wealthy landowners in designing their estates with the long-term future in mind also extended to others in more urban situations. London was growing rapidly in the mid-nineteenth century, and this was of concern to a small group of people who actively sought to improve the conditions for the workers and poor of the city. The countryside was then relatively close, so the idea of protecting areas as 'green lungs' for London that were accessible at weekends for those living in the city was born. From this idea, the policy of having a green belt around major cities in England has grown. Around London there are some important places for veteran trees and wood pasture such as Epping Forest and Burnham Beeches, which were preserved at the very beginning of this policy in 1878 by the City of London Corporation who took through Acts of Parliament to allow them to acquire these areas in perpetuity for the people of London [23] and has since been added to over the intervening years.

3.2.4 Conservation organisations

Amongst those responsible for drawing up the City of London Acts were some who had aspirations for something similar that had a more nationwide remit. Out of this the National Trust was born, an NGO that acquires land, as well as buildings, to protect them in perpetuity for the nation. The National Trust now owns some very special wood pastures in South East England like Hatfield Forest. Since then a variety of NGOs have arisen in the conservation sector, coupled with an increasing awareness of the importance of conservation generally which gained traction in the 1970s and 1980s in the UK, and this has helped protect wood pastures and their trees. The conservation sector in the UK is particularly complex with nature reserves owned and managed by a variety of governmental and non-governmental organisations as well as private landowners. Several important wood pastures in South East England are designated as SAC sites (such as Windsor Great Park, Epping Forest and Burnham Beeches), and others are protected in the National Law as Sites of Special Scientific Interest (the basic nature reserve designation in England) and sometimes also the National Nature Reserves, with examples being Ashtead Common and Hatfield Forest.

3.3 Problems that the trees face now

3.3.1 Lack of management and abandonment of grazing

The most common problem facing our old trees today is the abandonment of grazing and/or lack of management. With the abandonment of grazing, the older trees become heavily shaded by younger trees which have grown up as a consequence of lack of grazing [3]. Sometimes, the younger trees may overtop the older ones or grow through the canopy resulting in significant shade, which in the end may lead to an early death of the old trees. In places, once open wood pastures have also been planted with conifers which results in even faster losses.

Another common problem is the abandonment of management of the trees themselves, primarily a lack of regular pollarding. The majority of wood pasture sites in the South of England contain old pollards that may not have been cut for at least 100 years. This has led to large, heavy branches growing on often hollow stems, which have a high tendency to fall apart [3, 4].

3.3.2 Agricultural intensification

Many of the wood pastures in South East England were lost as a consequence of agricultural intensification. Sometimes, the trees remain (**Figure 9**), but the grasslands in between have been ploughed up, resulting in significant root damage. Sometimes, the grassland in parklands (see above) is ploughed, fertilised and reseeded, which as well as being detrimental for the trees also removes the bushes and other nectar sources, which are a natural part of the mosaic in a wood pasture system. Agricultural intensification also resulted in grubbing out of hedgerows to enlarge fields. The hedgerows often contained old trees, which may once have been part of an old wood pasture or common.

According to Rackham [10], the Georgian and Victorian eras saw the greatest destruction of wood pasture systems, as they were seen to have no further use. In the twentieth century, less was lost, but coniferisation of wood pasture sites was common, destroying the essential elements of the habitat. Britain is still thought to have a large proportion of wood pasture; however, there are no reliable statistics on the extent of the overall cover nor on its loss or fragmentation.

3.3.3 Urbanisation/planning

Increased pressure on the landscape for development has also had an impact on wood pastures. There are many examples of housing developments which have been built on old wood pastures. Sometimes, the trees are retained (**Figure 10**), but often this leads to conflicts of interest between the homeowners and the trees, due to concerns about safety or problems with roots. Often the space retained is not enough for the long-term good health of the trees.

3.3.4 Fragmentation

As a consequence of the above problems, wood pastures and parklands often now exist as isolated fragments in an otherwise intensively farmed or urbanised landscape. As the fragments become smaller and further apart, species dependent on the trees and their surrounding environment are less able to move between them, and their populations eventually become unviable. This is a particular problem in South East England as huge amounts of development and house building take place in the area around London.

3.3.5 Loss of traditional knowledge and skills

People are nowadays much less connected to the land, and many of the skills that were previously passed down through the generations have been lost. Farming methods have changed to be more mechanised, and the manual skills are no longer needed. There is really no first-hand knowledge of how the trees were pollarded, and most were last cut over 100 years ago, so there is no older

Figure 9.
Old Quercus *sp. tree that has been ploughed close to it in order to cultivate for arable crops.*

Figure 10.
Old Quercus *sp. tree that was part of a wood pasture on a wooded common but is now within a housing estate. Aspal close, Suffolk.*

generation of workers to talk to. As the work was once considered to be so commonplace, details were not generally written down, so it is necessary to (re)learn from other places, try to interpret from the structure of the trees or discover again from the beginning. In addition, the lapse in cutting of the trees means that they are now out of a regular pollarding cycle, something that our ancestors did not have to deal with, so we have to learn new skills to manage these trees. The consequences of our actions may take many years to become evident, so this is not a process that should be hurried.

Loss of skills also applies to animal husbandry. In the past domestic livestock were herded, and thus their grazing and browsing were directed. Today animals are put into fields unattended for most of the day. Some of the older traditional livestock breeds that developed in these systems are no longer available for use or are difficult to obtain, and the specialist knowledge for managing them is known by a small number of, often amateur, breeders. The lack of connection with the land also means that people are not used to coming into contact with grazing animals and may also see fencing as a restriction on their rights to walk where they like. This is especially true for the wood pastures in South East England, where visitor pressure is high and concentrated. Conflicts between visitors and livestock are common, such as with dogs.

3.3.6 Pests and diseases

Many of our old trees in wood pasture landscapes are suffering from new pests and diseases. The number of pests and diseases on trees has increased tenfold over the last 2 decades [31]. Some examples which are having an impact on the wood pasture landscape of South East England include oak processionary moth (Richmond Park, Ashtead Common), acute oak decline (Ashtead Common, Aspal Close) and ash dieback (Hatfield Forest) to name but a few.

4. Solutions

4.1 Learning to recognise the importance of the trees/landscapes

Much work is being done to raise the profile of wood pasture, which until the last couple of decades has had very little attention, particularly in nature conservation. The Ancient Tree Forum, Natural England and other partners through the Habitat Action Plan Group and the Ancient Tree Hunt have helped raise the profile of the ancient trees in wood pasture landscapes. In addition, these organisations have also worked hard to help share the knowledge regarding the management of these important sites and their old trees [3, 4].

4.2 Reinstate traditional management

For many wood pastures, the restoration and reintroduction of grazing animals is the most important management option for conserving these valuable habitats. This is potentially more important than any work which may be required on the individual trees. Restoring wood pastures needs to be carried out in stages, to make sure that the old trees are not suddenly shocked by fast changes to their environment [3, 4]. Removing competing trees should first involve removing the young

Figure 11.
View looking up from a tree that has been halo cleared to show the gap between its foliage and that of the surrounding trees—The first step in giving an old tree more light (Ashtead common).

Figure 12.
Traditional English longhorn cattle grazing a wood pasture at Cranborne chase, Windsor. These cattle are good at browsing and thriving on poor quality vegetation.

trees that grow under and through the canopy of the old trees, ensuring there is also a ring of a few metres around the canopy that is open (**Figure 11**). A few years later, the next phase can create a larger ring around the old tree, ensuring that the tree is not too exposed to the sun and wind. Doing this work by focusing on trees in groups is more effective and leads to better results [32] than only on individual trees.

Grazing may then be reintroduced in stages [33]. In the South of England, hardy sheep breeds have been used as an interim restoration phase for several years to help restore a grassland with greater proportion of herbs rather than woody, unpalatable species. This has then been followed up by the introduction of cattle. The use of traditional breeds of livestock (**Figure 12**) that are better at browsing and thriving in areas with poor quality fodder has been promoted by organisations like the Rare Breeds Survival Trust and the Grazing Animals Project. As a result, they are more commonplace in wood pastures now, and there is increasing experience in how to manage them.

It is unlikely that in most wood pastures, the clock will be turned back so that they will be managed as they were at a particular period in the past, nor is this probably desirable, but traditional methods are being used to inform future management.

4.3 Learn new skills

The management of lapsed pollards, well out of a regular cycle of cutting, is something that our ancestors did not have to deal with. Probably any particularly difficult trees were left uncut, and any that died would not have been a major concern as new ones could be planted and subsequently cut. Nowadays our lapsed pollards are precious because they are remnants of a landscape and culture that are no longer active and the numbers of trees are substantially lower than in the past and declining. We are aware of the value of these trees for biodiversity and aim to keep them alive as long as possible. This has required honing new skills in dealing with these trees, and early experiences in the 1980s and 1990s were not always positive.

From our experience of managing the lapsed pollards in wood pastures in South East England we have identified some actions which quickly result in the death or long-term decline of these trees, and others that currently appear to be more successful in keeping them alive; thus, the following recommendations can be made (see also [34, 35]):

1. Do the trees have sufficient light? Many older trees have lower canopies and are perhaps growing less vigorously than younger trees. Alternatively, the branches may have been drawn up because of the lack of light so that all the foliage is at the tips of the branches. The solution is to clear round the trees so that there is firstly a small gap between the tree and any neighbouring branches (see also Section 4.2) by removing young regrowth growing through the canopy.

2. Are the tree roots OK? Avoiding damage to the roots and the soil that they grow in from agricultural activities, building or other development or visitor pressure is very important. Although trees can be fenced, other solutions that are less visually intrusive include: re-siting agricultural activities (water trough, etc.), using a dead hedge, encouraging rough vegetation to grow around the tree or mulching (using a shallow layer of partially composted wood chip to protect the roots). For feature trees or those with lots of visitors, a boardwalk can be a solution.

3. Does the tree need pruning to stop it falling apart or falling over? We suggest that cutting a tree is the last resort, but for some top-heavy and fragile lapsed pollards, it may be the only option. Pruning should aim to remove as little leaf area and branch material as possible and to make wounds that are as small as possible (**Figure 13**). Cutting lapsed pollards as if they were still in a regular cycle of cutting is not generally successful although sometimes it can appear so initially. The growth that arises on the sides of large wounds close to the bolling eventually becomes very vulnerable as the branches get heavier and the decay resulting from the cut gets more extensive. Eventually, when this fails, it has the potential to tear away parts of the trunk of the tree as well as destroying the bolling.

4.4 New markets for products

As functioning wood pasture systems with managed trees and grazing livestock start to become a part of the landscape again, they need to become relevant and cost-effective to survive. Many are being restored because they are nature reserves, but in times of austerity and agricultural uncertainty (e.g. Brexit), the more sustainable they are, the more likely they are to survive. Wood products like firewood, bean poles, wood chip and charcoal can be produced from the trees along with meat from the grazing livestock (**Figure 14**). In recent years there has been a resurgence in interest in local and artisan products as well as improvements in marketing so that people are aware of the importance of buying locally. Local/regional festivals, farmers' markets, meat box schemes and site-based events, for example, the wood fest at Hatfield Forest, are methods by which it is possible to sell products at premium prices, and this can be achievable with the relative wealth in this part of England.

4.5 Agri-environment and other direct funding

It is possible to obtain financial support for wood pasture restoration and management in England at the moment. Many nature reserves are owned and managed by charities that gain funding from their members and grants via a variety of sources. In recent years this has included grants from lottery income and from taxes taken from companies that carry out, for example, landfill or quarrying activities as well as private funds. Agri-environment funding, i.e. from CAP, has been possible for wood pastures for many years. Currently there are area-based payments for both the restoration and the ongoing management of wood pasture as well as payments for working on veteran trees and the creation of decaying wood habitats on trees (amongst other potential payments that might be relevant to wood pastures), but some of these are

Figure 13.
Gradual reduction on a lapsed Fagus sylvatica *pollard at Burnham Beeches.*

Figure 14.
Tamworth pig and deer at Knepp Castle estate. Pork, ham and venison contribute income for the estate.

only accessible on high conservation value and/or large sites, and thus smaller areas may be less well supported. In addition, the administrative burden and obligatory requirements of the scheme mean that it is not attractive to all landowners.

4.6 Landscape-scale approach

In 2010 a report was published [36], commissioned by the Government, which was the result of a review of how England's wildlife and ecological network could be improved to help nature thrive in the face of climate change and other pressures (The Lawton Review). This concluded that England's wildlife sites, despite their diversity, did not currently comprise a coherent and resilient ecological network and therefore were not capable of coping with the challenge.

The report highlighted that the fragmented nature of the landscape, caused by human activity like development and intensive agriculture, when subject to rapid change like climate change, is unable to adapt quickly enough as species are unable to move through it. It thus proposed that a landscape-scale approach is needed to reverse fragmentation and environmental degradation. Wood pastures have the potential to play a key role in this due to their mix of trees and open areas. They are also potentially productive areas for both livestock (food production) and wood as well as being good for nature conservation.

The Lawton Review sets out a clear vision with a change of direction from the current emphasis on wildlife in isolated reserves and towards whole landscapes that are vibrant, wildlife-rich and ecologically functioning. To do this, Lawton says that we need to make our network of sites bigger, better and more joined up. This call for action has been supported through various funding opportunities and will hopefully benefit wood pasture and promote low-intensity management of larger areas.

4.7 Rewilding

Rewilding is a concept that varies in terms of what it means across Europe. In some places however, it is proving to be an interesting and viable management regime for wood pasture. Areas where rewilding has been tried are not necessarily those that have a long history as wood pasture nor those where there are particularly important tree populations, but this does not stop them from contributing to a landscape-scale approach.

Case study: The Knepp Estate in Sussex took the decision in 2001 to stop traditional farming and remove all the internal fences and boundaries. Several different species of livestock (*Dama dama*, longhorn cattle, Tamworth pigs and Exmoor ponies, **Figure 15**) were introduced and allowed to roam freely. This allows the animals to decide where they want to feed, wallow, calve and sleep. The idea builds on the concepts and ideas of Frans Vera and his book *Grazing Ecology and Forest History* in 2000 [6]. The animals are allowed to drive the system, and the result so far after some 20 years is a dynamic wood pasture system, where biodiversity is thriving [37].

Figure 15.
Exmoor ponies used as part of the rewilding project at Knepp Castle estate.

DOI: http://dx.doi.org/10.5772/intechopen.86100

In addition, the Knepp Estate has been able to sustain itself financially by selling the meat from the animals and developing agro-tourism, along with receiving agri-environmental subsidies and carrying out other farm diversification activities [38].

4.8 Create decaying wood habitats including new pollards

The key elements in wood pasture that are so important for biodiversity are the old trees and the decaying wood they contain and produce. Decaying wood provides conditions that are suitable for a wide range of species (especially fungi and insects); many of these species are very rare, and some have very precise requirements in terms of decay type and moisture level. Some species may also have difficulty colonising new sites if they are too far apart. As described above, many wood pasture sites are fragmented with few old trees remaining, and there may be a large generation gap between the old trees and their successors.

Sometimes it is desirable to attempt to create some decaying wood habitats to help close up this generation gap or provide habitats where there are none currently. Pollarding is one reliable technique, whereby through regular cutting, the trees develop hollows at a younger age than trees that have not been pollarded [39]. Other techniques, to create decaying wood habitats in younger trees, known as veteranisation [40], involves damaging younger trees in a variety of ways by mimicking nature. The principle is never to do this to the old trees but to those that need to be felled or cut for other management reasons. A variety of different techniques have been attempted such as the creation of woodpecker-like holes using a chainsaw (**Figure 16**), making bird nesting boxes within the trunks of sound trees and mimicking bark damage by horses or sheep at the base of the trees using chainsaws or axes; there is plenty of scope for experimentation.

4.9 Protection/reinstatement of hedges

In agricultural systems where scattered trees in the hedges are acting as a different form of wood pasture, a big problem in the past has been the removal of hedges to make larger fields allowing easier cultivation with larger machinery. Any trees in the hedges were generally lost too and were rarely incorporated into the fields. Hedges are now protected in England, and there are grants for replanting and managing them. In highly productive areas, this has been an important way to increase the density of trees in the landscape along with the scrub species in the hedges.

Figure 16.
(A) Woodpecker hole created with a chainsaw at the bottom and holes created by a woodpecker above it, a few years after the work was done. (B) Newly created Quercus *pollard.*

5. Conclusion

The wood pastures of the South East England contain some of the highest biodiversity remaining in an otherwise heavily exploited landscape. There are however significant challenges including conserving the existing protected sites, identifying other valuable sites that are not protected, creating new sites for the future and linking and buffering these areas. New uses for the trees in combination with grazing, to make wood pastures both attractive and sustainable, are perhaps the way forwards if we are to ensure that there are wood pastures with old trees into the future.

Acknowledgements

We are grateful to Paul Read for the use of his photograph of trees at Goswold Hall and his comments on the section on the history of field systems. We thank Ted Green for his tireless work raising awareness of the value of old trees and wood pastures in the UK and who, together with our other colleagues in the Ancient Tree Forum, has helped shaped our views over the last 30 years.

Author details

Helen Read[1*] and Vikki Bengtsson[2]

1 City of London Corporation, Burnham Beeches, UK

2 Pro Natura, Alingsås, Sweden

*Address all correspondence to: helen.read@cityoflondon.gov.uk

References

[1] Perry S. A strategic view of the issues for wood-pasture and parkland conservation in England. In: Rotherham ID, editor. Trees, Forested Landscapes and Grazing Animals. London/New York: Routledge; 2013. pp. 356-375

[2] Peoples Trust for Endangered Species. UK Wood Pasture & Parkland Network [Internet]. Available from: https://ptes.org/wppn/videos-links-downloads/ [Accessed: September 9, 2018]

[3] Read HJ. Veteran Trees: A Guide to Good Management. Peterborough: English Nature; 2000. 176p

[4] Lonsdale D. Ancient and Other Veteran Trees: Further Guidance on Management. London: The Tree Council; 2013. 202p

[5] Vetree—Veteran Tree Network. Vetree Veteran Tree Network [Internet]. Available from: www.vetree.eu [Accessed: September 9, 2018]

[6] Vera FWM. Grazing Ecology and Forest History. Wallingford: CABI; 2000. 582p

[7] JNCC. UK biodiversity Action Plan Priority Habitat Descriptions - Wood pasture and Parkland. Peterborough: JNCC; 2011. 5p

[8] Godwin H. The History of the British Flora: A Factual Basis for Phytogeography. 2nd ed. Cambridge: CUP; 1975. 541p

[9] Kirby K, Watkins C. The forest landscape before farming. In: Kirby K, Watkins C, editors. Europe's Changing Woods and Forests. Wallingford: CABI; 2015. pp. 33-45

[10] Rackham O. Ancient Woodland. Kirkcudbrightshire: Castlepoint Press; 2003. 584p

[11] Hooke D. Trees in Anglo-Saxon England. Woodbridge: The Boydell Press; 2010. 310p

[12] Rackham O. Trees and Woodland in the British Landscape. London: J.M. Dent; 1976. 204p

[13] Barnes G, Williamson T. Ancient Trees in the Landscape Norfolk's Arboreal Heritage. Oxford: Oxbow Books; 2011. 179p

[14] Peterken G. Woodland History in the British Isles—An interaction of environmental and cultural forces. In: Kirby K, Watkins C, editors. Europe's Changing Woods and Forests. Wallingford: CABI; 2015. pp. 265-278

[15] https://www.oss.org.uk/what-we-do/commons/ [Accessed: February 09, 2019]

[16] Hooke D. Early wood commons and beyond. In: Rotherham I, Agnoletti M, Handley C, editors. The End of Tradition? Sheffield: Wildtrack Publishing; 2014. pp. 107-120

[17] https://www.cityoflondon.gov.uk/things-to-do/green-spaces/city-commons/ashtead-common/Documents/Ashtead-common-local-plan.pdf [Accessed: February 09, 2019]

[18] Roberts J. Royal Landscape. The Gardens and Parks of Windsor. Yale: Yale University; 1997. 596p

[19] Alexander K, Green EE. The nature conservation work of the Crown Estate in Windsor Forest and Great Park. British Wildlife. 2013;**24**(5):305-315

[20] Rackham O. The Last Forest. The Story of Hatfield Forest. London: J.M. Dent & Sons; 1989. p. 302. 3560p

[21] Glimmerveen I. The future potential of wood pastures. In: Rotherham ID,

editor. Trees, Forested Landscapes and Grazing Animals. London/New York: Routledge; 2013. pp. 339-355

[22] Tusser T. Five Hundred Points of Good Husbandry. 1984th ed. Oxford: Oxford University Press; 1580. 344p

[23] Dagley J, Burman P. The management of the pollards of Epping Forest: Its history and revival. In: Read HJ, editor. Pollard and Veteran Tree Management II. London: The Richmond Publishing Co. Ltd; 1996. pp. 29-41

[24] Green T. Ancient trees and wood pastures. In: Rotherham ID, editor. Trees, Forested Landscapes and Gazing Animals. London/New York: Routledge; 2013. pp. 127-142

[25] Le Sueur ADC. Burnham Beeches. A study of pollards. Quarterly Journal of Forestry. 1931;**1931**:1-25

[26] Barr C, Howard B, Bunce B, Gillespie M, Hallam C. Changes in Hedgerows in Britain between 1984 and 1990. Huntingdon: Institute of Terrestrial Ecology; 1991

[27] Barr C, Gillespie M, Howard D. Hedgerow Survey 1993. Institute of Terrestrial Ecology: Huntingdon; 1994

[28] Barr CJ, Gillespie MK. Estimating hedgerow length and pattern characteristics in Great Britain using countryside survey data. Journal of Environmental Management. 2000;**60**:23-32

[29] Farjon A. Ancient Oaks in the English Landscape. Surrey: Kew; 2017. 348p

[30] Abrego N, Christensen M, Bassler C, Ainsworth M, Heilmann-Clausson J. Understanding the distribution of wood-inhabiting fungi in European beech reserves from species-specific habitat models. Fungal Ecology. 2017;**27**(B):168-174

[31] Forestry Commission. Forestry Commission Pests and Diseases [Internet]. Available from: www.forestry.gov.uk/pestsanddieseases [Accessed: September 9, 2018]

[32] Alexander K, Stickler D, Green EE. Is the practice of haloing successful in promoting extended life?—A preliminary investigation of the response of veteran oak and beech trees to increased light levels in Windsor Forest. Quarterly Journal of Forestry. 2010;**104**(4):257-265

[33] Williams R, Read H. Rare breeds and public access—Do they mix? Enact. 1997;**5**(4):12-14

[34] Read HJ, Wheater CP, Forbes V, Young J. The current status of ancient pollard beech trees at Burnham Beeches and evaluation of recent restoration techniques. Quarterly Journal of Forestry. 2010;**104**(2):109-120

[35] Read HJ, Dagley J, Elosegui J-M, Sicilia A, Wheater CP. Restoration of lapsed beech pollards: Evaluation of techniques and guidance for future work. Arboricultural Journal. 2013;**35**(2):74-90

[36] Lawton JH, Brotherton PNM, Brown VK, Elphick C, Fitter AH, Forshaw J, et al. Making Space for Nature: A Review of England's Wildlife Sites and Ecological Network. London: DEFRA; 2010

[37] King M. The Knepp Vera Conference: The case for creating new wood pastures. British Wildlife. 2017;**29**(1):27-33

[38] Marren P. The great rewilding experiment at Knepp Castle. British Wildlife. 2016;**27**(5):333-339l

[39] Sebek P, Altman J, Platek M, Cizek L. Is active management the key to the conservation of Saproxylic biodiversity? Pollarding promotes the

formation of tree hollows. PLoS One. 2013;**8**(3):e60456. DOI: 10.1371/journal.pone.0060456

[40] Bengtsson V, Niklasson M, Hedin J. Tree veteranisation. Using tools instead of time. Conservation Land Management. 2015;**Summer**:14-17

Chapter 2

The Archeology of Woodland Ecology: Reconstructing Past Woodmanship Practices of Wooded Pasture Systems in Italy

Diego Moreno, Roberta Cevasco, Valentina Pescini and Nicola Gabellieri

Abstract

Recently, the woodmanship practices and localized naturalistic knowledge have been acquired as indispensable for the European forest heritage conservation as well as for restoration of individual woodland landscapes. Minor importance has been given to the historical approach, both in the study of the local resources ecology and of the local societal context. Using the results of a series of case studies—applied to the knowledge and planning of sites that host present (or have hosted in past) wooded pasture systems and their environmental legacy—the chapter shows the interest of the environmental resource archeology (ERA): a "multisource approach" in reconstructing past management systems practices and underpinned lore. ERA is inspired by the method and sources of the English historical ecology and topographical history employing both archives and field evidences (palynology, anthracology, etc.).

Keywords: environmental resource archeology, historical ecology, wooded pasture/meadow systems, multi-proxy approach, NW-Central Italy

1. Introduction

In this chapter, we explore the potential of historical approaches to the study of the ecology of present woodland ecosystems in order to identify the local practices involved in the past management. In particular, an archeological and multisource reconstruction of woodmanship [1] practices and their associated local environmental knowledge is proposed. The title "The Archaeology of Woodland Ecology" explicitly relates to a conference held in 1982 by the British Environmental Archeology Association, which is considered as one of the main starting points in environmental archeological studies in UK (and in Europe). Our interest for such an early milestone relates to the attention that the conference holders had demonstrated to the historical ecology approach and to the studies of Oliver Rackham on the Mediterranean landscapes of Crete [2, 3]. However, Rackham's historical approach played little or no part in the subsequent years of development in environmental archeology. In fact, only relatively recent contributions by environmental archeologists [4] have finally fully acknowledged

Rackham's use of multiple sources as well as the regressive historical approach. This achievement appears in contrast to the prevailing quantitative and structural ecological models incorporated from paleoecology into environmental archeology studies. Following these lines, in the next pages, we will confront the main topic of the first edition of the Journal of Environmental Archeology ("Fodder: Archaeological, Historical and Ethnographic studies," 1998), discussing this theme from the perspective of Rackham's historical ecology or, in other words, relating our experiences in "Environmental Resource Archaeology." Environmental Archeology (EA) commonly refers to an archeological–paleoecological approach to the study of the "human paleoenvironment": the analysis focuses on the relationship between human activity (and behavior) and the environment (considered as natural ecosystem) in the past [5]. Instead, historical ecology studies from the British tradition have proposed a different historical approach, considering the vegetation canopy of a given site as a particular kind of artifact. Consequently, environmental resources are considered as "social products" ("part of local society") and their ecology appears historically defined by activation and production practices, conditioned by conflicts of control and access in local communities, and by local environmental knowledge [6, 7]. In contrast to a purely environmental study, we adopt an historical and archeological analysis aiming to extrapolate localized production and consumption systems and the related social practices that have over time determined the features of the resource and their unique environmental state. This approach, named "Environmental Resource Archaeology" (ERA), has been developed in the last 30 years by the LASA team (Laboratory of Environmental Archeology and History) at the University of Genoa in different projects and applications in North West and Central Italy. The development of ERA methodological approaches is linked with particular microhistorical geographical approaches [8, 9]. "History" is purposely defined as any historical process that links actions and practices, and their material effects on resource ecologies, avoiding historical economic generalizations in term of civilization (e.g., the "longue durée" historical model proposed by the "géohistoire" approach of Annales School and F. Braudel) [10, 11]. Similarly, the ERA approach avoid the simplification and generalisation of the local social context as produced by some branches of the "socio-cultural history" in terms of tradition/traditional sociability; for instance, the recentely-coined definition of "bio-cultural heritage", suggested by conservation studies, needs to be further clarifyed analysing the concrete relationships between societies and environment, and paying attention to the form and content of these relations (ownership, access, knowledge, etc.) (e.g. [12, 13]). Focusing on the environmental characterization and reconstitution of local practices, we attain a high societal/economical resolution. For example, the biodiversity of the present site is considered a marker of bio-diversification processes that consistently affect the environmental systems. However, the analysis of small-scale biodiversity variation effects registered in palynological/anthracological sources when contrasted with textual, cartographic, or iconographic sources available at the site of observation scale, sheds new light on the historical dimensions of bio-diversification itself [14]. It is important to recall—perhaps for all non-historian readers—two main principles in the use of evidence as acknowledged by two different but convergent historiographical school: those of English local history and the Italian microhistory. Both traditions are interested in the historical processes that led to the production of the present locality and site "characters" ("characters" is intended as historical products/issues) as well as of the material features of the landscape [15–17]. The rapidly growing interest in an ever widening range of disciplines due to the existence of past multiple uses in the management of forest resources has been largely generated by the growing problems of identification and conservation of biodiversity, both on a global (e.g., studies on land bearing trees of savannah-type cfr. [18]) and local individual rural landscape scale [19].

In Italy, it becomes evident, at a site observation scale, that the loss of biodiversity caused by the expansion of secondary forest formations on previous agricultural land (field, meadows, etc.) is the same loss effects that follows the abandonment (or lack of management) of many woodland type (e.g., coppiced woodlands or abandoned wooded grasslands). The regressive historical approach takes the previous ecology of the abandoned (or re-naturalized) resources as reference. This historical confrontation—in order to reconstruct the past biodiversity of the site—is seldom taken in account in paleocological studies. Paleoecologist are rather interested in global climatic continuity/discontinuity, and rarely explore the topographical scale of the observed site both in space and time. Generally, material historical processes are surrogated by a more or less linear coevolution model of "Man and Nature" relationship. On the other hand, in historical geographical studies, wooded pasture is still assumed as "semi-natural system" rather than a fully historical issue of past sylvicultural/woodmanship [20]. Consequently, dealing with environmental conservation policies, they are prone to assume global bio-cultural heritage and rarity values or traditional cultural value-oriented policy [13, 21]. The reconstitution of past biodiversity at a landscape scale would of course be the main goal of environmental history and archeology. Nevertheless, at least in Italy, the landscape archeology seems to be more focused on settlement than on environmental resources [22]. Any confrontation or interdisciplinary project involving such different disciplinary participation is possible without taking into account the historical nature of the observed processes. By such a premise, it became necessary to question what kind of historical approach and sources have a scientific ground in the application of historical ecology researches. Has such a historical approach been applied, for instance, in the bio-cultural heritage definition as proposed by the conservation organizations [19]?

The ecology of past multiple systems (and in particular of wooded pasture systems) offers a very interesting interaction between forest resources, their biodiversity, and the actions and practices that has controlled and generated their environmental diversity. The role of traditional knowledge in shaping forests and their biodiversity in Europe and across the world has already been stressed in several publications [21, 23] drawing attention to differences in the approaches of conventional historical researchers nurtured by global environmental history, as opposed to historical ecology. In the former, the bio-diversification processes (as historical processes) are largely neglected or subsumed into general observations concerning global change, or are embedded in presumed "a-historical" and "traditional" economies and standing agro-sylvo-pastoral systems. Instead, the latter proposes that in the field of environmental and cultural conservation studies, such broad assessments are necessary prior to multi- or interdisciplinary applications.

Bio-diversification processes may be observed in, for example, paleontological and paleoecological studies at very different timescales. In the historical ecology case, diversification changes are observed at an historical time scale. As consequence, bio-diversification processes can be addressed through specific historical and historiographical topics (for an in-depth discussion of the issues emerging from studies of bio-diversification processes see [14]).

The case studies selected in this chapter (**Figure 1**) have been explicitly developed in different applied research projects devoted to specific sites and individual landscapes starting from the multisource methodological premises of the British historical ecology.

The historical approach is adopted here in order to identify and characterize past and ancient wooded pasture ecology. Also, it addresses more theoretical problems as posed by the historical ecology research development. In our case studies, woodland ecosystems are assumed as ecosystems concerning

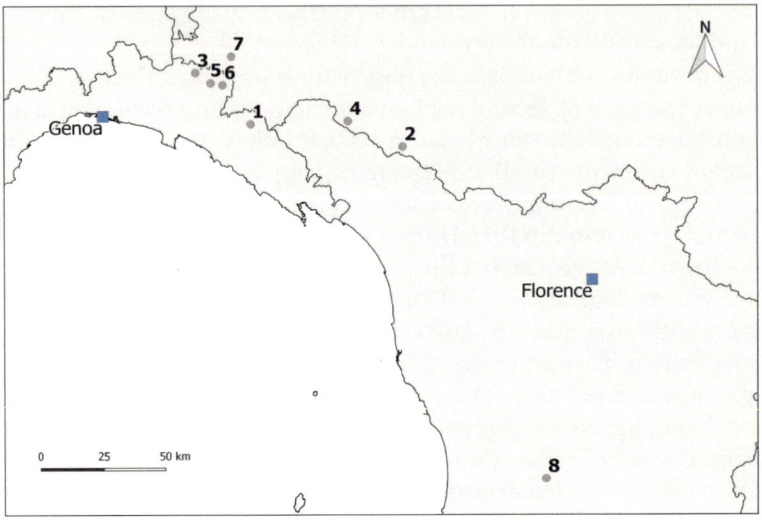

Figure 1.
Case studies location map. (1) Upper Vara Valley; (2) Prati di Sara; (3) Montarlone—Pian Brogione; (4) Prato Spilla; (5) Mogge di Ertola—Moglia di Casanova; (6) Upper Trebbia-Aveto Basin; (7) Ceno—Nure valleys; and (8) Montieri.

land-bearing trees showing savannah-type ecologies (sensu [24, 25]). Each woodland that will be considered here appears to be composed of pure or mixed populations of silver fir (*Abies alba*), beech (*Fagus sylvatica*), and Turkey oak (*Quercus cerris*). Moreover, this section is also intended to contribute to a better evaluation of the sources (textual, cartographical, sedimentary, anthracological, oral, etc.) and purposes when such an analytical historical approach is adopted in geographical studies on woodland landscapes and resources. Field and documentary evidence collected during multidisciplinary historical ecology studies of sites in the northern Apennines of Liguria and Tuscany will be commented upon. This topographical scale research indicates that the key drivers in associated biodiversity changes resulted from medieval and post-medieval transformations in management practices.

The case studies are arranged in order to underlay the importance of the topographical and social local context (e.g., exploring local forest resource terminology meaning). Such an analysis is possible when the language of textual sources is deciphered in the light of the local taxonomy system, as is the case of the 1822 Forest Inventory in Upper Vara Valley (paragraph 2). Field and sedimentary evidences are contrasted focusing on the historical ecology of beech and fir wooded pasture sites and their bio-diversification processes in the Upper Trebbia and Aveto Valleys (paragraphs 3–4). Finally, the recent development of environmental resource archeology related to the study of charcoal hearth soils (paragraph 5) demonstrate the relevance of anthracology in woodmanship practices characterization and in the comparison of wooded pasture system at a European scale.

In terms of final remarks, the case studies suggest that historical ecology approaches, applied locally, can raise key questions and new data, which can differ from those of the traditional archival and textual based studies, as well as those from non-historical ecological research. The methodology adopted in a locally based approach should use specific historical analyses, documentary, and archival sources, together with archeological and sedimentary evidence. This results can contribute not only to finding answers to the key questions, but also to the generation of new research directions on forest resources ecology and their applications in the European rural heritage identification.

2. Wooded pasture: looking for the local historical context

In 1822—at the very end of the customary regime—the entire forest resources of a high valley of the Eastern Ligurian Apennines (Upper Vara Valley) had been described orally in front of the local judge. Excluding the chestnut orchard—considered at the time as a purely agricultural land use type—each landowner has been questioned about the rights and management system of their land. These descriptions (called "Consegne") were transcribed by a notary chancellor, had a legal value, and were collected in the agenda for the formation of a Forest Inventory in charge to the Forest Administration of the Kingdom of Sardinia. The *Consegne* are textual source type that in recent years have provided historical ecology research in the NW Apennines with a reliable sample of hundreds of wooded parcels with the opportunity to localize them and assess the precise environmental content of the site [26, 27]. As commonly attested in this type of documents, the language of the oral answers interacts poorly and confusingly with the form and terms proposed by the administration's questionnaire. Interestingly, the text of each single "parcel" description provides us an in-depth insight of the local environmental resources vernacular terminology. This last is a problem that renders this documentation virtually useless for the Forestry Administration which, in the following years, did not unify a coherent description of all the local woodlands documented in the Ligurian mountains. At the end of the enquiry, for the administrative/prescriptive purposes, the local land uses types were reduced to a few (4–6) conventional forest categories. The rich resources vernacular terminology was converted in the international economic and forestry scientific literature classes (largely written in French in the Kingdom of Sardinia) and already adopted by the administration before the initiation of the 1822 inventory. In a small sample, such as in the studied parishes of the Upper Val di Vara, it is possible to discover and characterize 21 different forms of local use of the forestry resources.

2.1 Local forms of use of the wooded soil: from text terminology to the taxonomy of local environmental knowledge

Starting from the apparently confusing contents of the answers—mixing local and administrative terms—it was nevertheless possible to identify various local forms of forest uses. By breaking down the precise references that each of the owners was obliged to elucidate in order to establish the arboreal species composition, management practices, and obtainable productions, it is possible to identify a global logic in the local system that appears in opposition to the logic of Forestry Administration. As demonstrated in previous works [28], local language terminology corresponded to a coherent local taxonomy of forest resources which had juridical power both in technical and legal terms during the customary regime [29]. The history of these mountain woods resources, as in the whole Mediterranean area, appears to be dominated up to these years by animal production economies; indeed, the forest resources' types appear to be largely placed in a variety of common land use systems. The commons (variously called "comunaglie," "comunelli" concerning this part of the Ligurian—Emilian—Tuscan mountain reliefs) are pivotal in the organization of seasonal movements of livestock. Until the abolition of the commons rights on land from the late nineteenth century to the beginning of twentieth century [30], they appear to have been located in the summer quarter ("alpi" dial.) of the significantly extended double facing—South Tyrrhenian littoral pasture and North Po Plain pasture—transhumance seasonal movements. Ovine and caprine flock (but also, in minor amount, bovine herds) lay during winter grazing in long fallow lands, marshes, and coastal lagoons (but also, in some specialized systems that link sheep and olive and vineyards cultivations) on the Tyrrhenian shore ("marine"/"maremme") [31] or conversely—north facing—on the Po Valley plain field and marshland ("valli" dial.) [32]. It has already been noted that the existence of different types of multiple use practices in common lands became the framework in the post-medieval age. This contributed to the characterization of a history of highly conflictual use of pastoral resources (developing the medieval practice of clan s' feuds) among the family kinships entitled to access (see the central role of common land in the microanalitycal essay of one of these Apennine valleys, in [33]). In the microhistorical-geographical analysis of this type of textual sources, the conflicting use of this technical terminology clearly emerges. A linguistic tension (and a consequent or ambiguous linguistic use of the terminology) is always present both outside and also within the social framework of rural communities in the production of discourses or texts concerning forest resources, land use and related production practices [16, 17]. For example, the two local woodland uses type, (cfr **Figure 2**, column 1) *Macchia* and *Selva*, are not intended as synonyms for high forest of beech, but are clearly distinguished in the 1822 text and are actually distinguishable when placed in different aspects or phases relating to the positioning of the single parcel alongside the local utilization cycle. This exercise of contextualization shows that a generic reference to the locality, or to an imagined local context, does not allow this documentary material to be treated as a product in a supposedly homogeneous social bloc (e.g., the rural community? The peasant knowledge?). Moreover, does not allow us (or let imagine) a linear, so-called "traditional transmission", of the local practices and related environmental knowledge as they appear mentioned in the text (e.g., assuming the existence of an a-historical traditional ecological knowledge).

2.2 Commons, the ubiquitous wooded pasture systems

As one can see from **Figure 2**, of the 21 local taxa referred to different wooded land use, only 4—named in the 1822 documents as: "Selva arborata," "macchia,"

UTILIZZO	SPECIE								FORME DI TRATTAMENTO				PRODUZIONI PRINCIPALI		
									FUSTAIA		NON-CEDUO*				
	Fs	Qc	Cs	Fo	Oc	Esp.	Ag	Psp.	T.SALTu.	TURNIa.	SGAMOLLO	CAPITOZZA	PASCOLO	FRONDA	ASSORT. LEGNAME
1 Selva arborata	*								S	100-150	▽	▲	◆	◇	□
2 Selva arborata	*	*							S	200	▽		◆	◇	□
3 Macchia	*								?	?	▽		◆	◇	?
4 Bosco arborato	*				*						▽.		◆	◇	
5 Bosco arborato	*								S	100	▽		◆	◇	□
6 Boscaglia arborata	*										▽	▲	◆	◇	
7 Bosco arborato	*	*							S	100-200		▲			□
8 Terra pascolativa	*												◆		
9 Terra arborata	*										▽	▲	◆		
10 Selva arborata		*									▽	▲	?	?	
11 Bosco arborato		*							?				◆	Frutto?	
12 Bosco arborato	*	*	*										◆	◇	
13 Bosco di		*		*								▲	◆	◇	
14 Bosco di			*		*						▽		◆	◇	
15 Boscaglia di		*				*					▽	▲	◆	◇	
16 Boscaglia di					*	*					▽		◆	◇	
17 Boscaglia di		*	*	*	*		*				▽	▲	◆	◇	
18 Terra arborata		*									▽		◆	◇	
19 Terra arborata di		*	*	*	*						▽		◆	◇	
20 Terra arborata		*					*	*			▽		◆	◇	
21 Terra arborata							*				▽		◆	◇	

Figure 2.
(Upper Vara Valley—SP) The woodland resources of the past Mandamento of Varese Ligure in 1822. Local land use type (in the first column of the figure), species composition (in second column), management practices (in the third column), and main production (in the fourth column) ([28], Table 35). Sources for the construction of the table were texts of the "consegne" (consignment) made to the Forestry Administration of the Kingdom of Sardinia of each individual forest parcels ("tenute di bosco") in observance of the law RR Patenti 13.7.1822.

"bosco arborato," and "boscaglia arborata"—produce wood assortments employing practices that characterized high-standard forest management [26]. These four wood types were composed of pure or mixed populations of beech and oak (especially Turkey oaks), each subjected to a cutting/harvesting practices; a selective cut of every chosen single tree with a cut cycle estimated by users of 100–200 years, called in Italian "taglio saltuario," and in French "jardinage." These types of high-standard forests occupied very limited areas in the parish common lands in comparison to the spread of all the others types of lands bearing trees. Largest extension was covered by wooded meadows and pasture devoted to fruit collection, pasture, hay making and, as well as bare areas, dedicated to the mixed production of hay and grazing (e.g., see the Upper Secchia Valley iconographic documentation). All of the 21 local taxa were subjected in 1822 to grazing rights, and even 17 are noted to specially produce leaf for fodder through different forms of treatment of the whole arboreal layer. These local land use types (column fourth central production of leaf fodder) consist of wooded meadow/pasture systems.

Figure 2 shows that in the Upper Vara Valley, the production of leaf fodder is almost entirely distributed in woods with pure or mixed beech composition (Fs = *Fagus sylvatica* in the second column of the table) and Turkey oak (Qc = *Quercus cerris*). Moreover, green pruning ("scalvo") practices were allowed in the management of high-growing stands ("fustaie", or selected areas of wooded pastures with beech or oak to produce fodder, acorns and timber, i.e. 1,2,5,7 in the first column of the figure). These are separately specified documenting the employment of shredding ("sgamollo") and pollarding ("capitozza") cuts. The production of leaf fodder and pasture is also reported in the local forms of wood use (first column no. 12, 14, 17, and 19) populated with chestnut trees, which practically formed different shapes of chestnut orchards (Cs = *Castanea sativa*) second column of the figure). It is interesting to note that the type bearing chestnut trees no. 17 appears mentioned as pollarded, although in the *Consegne* documentation the production of charcoal is never mentioned (for this correlation see paragraph 5). As we will see in following section, wooded pasture and meadows—as directly documented in the descriptions of many wooded parcels in 1822 in Vara Valley—are currently involved in the exploitation of individual, scattered small fields, setting a temporary form of minor cereal crop growing (based on the controlled fire practice called "ronco"). This temporary agriculture practice [26] must not be confused with the slash-and-burn agriculture systems, as frequently occurs in archeological and geographical literature. The right to set up the temporary field ("roncare, campeggiare") in the wooded commons is part of a regular cycle of cultivation treatments and practices, in which the crop itself is not the more important product; the main goal is to control the excess of organic substance and the invasions of shrubs and non-palatable herbaceous species due to intense grazing. In these mountains, such specialization has generated many local variations in the use of wooded pasture. This is also the case, for example, of the so-called "alnoculture system," involving a cyclical management for agricultural plots of small populations of gray alder (*Alnus incana*) (for an insight focus, see [34, 35]). Resulting from this discussion of the interconnection with wooded pasture systems, it is important to underline that—in the nearest Upper Aveto valley—in a slope in which "ronco" practices have been located by cartographic sources at least since 1720 AD—pollen analysis offers evidence that the "alnoculture" system appears present during the middle Ages on the slope previously occupied by fir wooded pastures [36, 37].

3. Past beech wooded meadows ecology (palynological, cartographic, and field sources)

In the Ligurian-Tuscan-Emilian portion of Apennines, it has already been noted that fragments of beech wooded meadows appear located in correspondence with sheep transhumance paths in activity up to the mid-twentieth century. The transhumant flocks that used the summer pastures ("alpi/alpeggi") were stationed (in winter) on the Tyrrhenian littoral pasture ("maremme") or in the Po plain marshland ("valli"). As previously outlined, the flock movements involved conflicts between the mountain communities and parishes. In this frame, litigation issues on the borders also arose between the state entities (the surroundings leading city-states of Parma, Reggio, Lucca) or the local lordships that had acquired and maintained a jurisdiction of the grazed area. Thanks to such disputes, in different archives (local and central), we can find series of manuscript maps that provide evidence of management practices that marked the beech wooded meadows resources during the post-medieval age. A fine example is a view map representing the head of the Upper Secchia Valley—in Tuscan-Emilian eastern portion of Apennine exactly at the sources of the Secchia river—that shows the vegetation cover of these slopes in detail. It has been produced by an anonymous painter-cartographer (or possibly a local land surveyor that seems—in drawing this map—to prolong a late medieval tradition of land representation). The map—conserved in the collections of the State Archive of Parma—is not dated, but can easily be attributed to the second half of the sixteenth century (we suggest 1590 ca.). This cartographic document can offer several details that will now be discussed. The different forms of treatment of beech and trees distribution (isolated, kept in small groups or rows) are recognizable. Land-bearing trees show similar management forms (or cutting cycle), as those interpreted using 1822 text on beech wooded meadows and pasture for the Upper Vara valley. In the water color drawing, it is possible to distinguish the presence of high-standard beech woodland concentrated in some stretches or patches (**Figure 3**) at the watershed level. Moreover, large portions of meadows and bare pastures appear where every single or small group of beech trees is represented showing a shape due to the different cutting practices of shredding, pollarding, and coppicing (**Figure 4**).

Shapes of the trees allow a comparison with the reconstruction of the woodmanship practices in medieval English wooded pastures [38] (**Figure 5**).

As recognized in recent reviews of Italian wooded pastures [23, 39], twentieth century forestry handbook writers and authors—although always favorable to the abolition of collective rights in the woods—positively acknowledged many of the environmental functions of the use of these cutting and harvesting practices. We find field evidence of the effects of activation on the whole ecosystem observing—at the site level—fragments of the landscape of the ancient wooded meadows and pastures. They were found on the same slopes of the Upper Secchia Valley painted in 1590 map, in a site called "Prati di Sara," in the summer of 1995 during a field survey aiming a vegetation map production [40]. The site extends over low-inclination slopes between Mount Cusna (2120 m asl), Le Borrelle (1623 m asl), and M. Bagioletto (1758 m asl) beyond the current beech forest limit (slightly above 1600 m asl). The site is located in a strip of subalpine vegetation that rises up to the highest peaks, characterized mainly by the blueberry moorland interspersed with alpine-like grassland. The botanical survey and measurements were carried out along a 20-m transect that passes through a group of beech trees and continues into the pasture in the dominant wind direction SSE-NNW. A summary of these first observations is presented in **Figure 6**.

It is interesting to note that while the geographical name meaning meadows ("prati") recalls the practice of mowing the herbaceous layer, it was actually still in

Figure 3.
Patches of high-standard beech woodland in a landscape of bare meadow-pasture (detail of the main watershed of Upper Secchia Valley in a Ms map dated 1590 ca A.S. Parma, Confini, b.205/7b, dis.887).

use here until the 1960s, as revealed from oral sources collected in the surroundings parishes in the early 1990s. Later, the effect of mowing has been substituted by ovine and bovine grazing. However, due to the gradual reduction of those herds, the environmental effects of grazing become thinner, until 1995 when our observation on the biodiversity of herbaceous layer were made.

At the time of our survey, we found (**Figure 6**): isolated trees (pollarded beeches) (a) that characterized the vegetation of the beech meadows-pastures with respect to the surrounding areas consisting of bare grassland (b). The wooded meadow-pasture occupied most of the flat areas in 1995, where it was probably easier for cattle to be stationed. On this site the presence of good forage grass (e.g., *Poa alpina*, *Phleum alpinum*, *Festuca rubra* and *Agrostis tenuis*) increases by one-third in areas bearing trees (a). On the contrary, the floristic composition of the bare meadow-pasture (b), where *Nardus stricta*, *Festuca ovina* and *Avenella flexuosa* prevail, is considerably impoverished.

The diagram in **Figure 6** represents the positive effects of the scattered pollarded beech trees on the herbaceous vegetation of the wooded meadow-pasture as recorded along a 20-m transect. In the area where the beech trees grow (a), the biomass of the herbs layer increases (1) (the height has been measured as an index) and the fodder ratio (3) (the ratio established between the presence of good fodder species and the total number of species detected) is tripled. In the figure, the graph has been duplicated in order to suggest the possible multiplication of the positive effects on a wider surface subjected to the beech grassland-pasture-system in which the single pollards showed a regular spacing of 20–22 m. This last measure is obtained comparing from the "fossil parcel" of beech wooded meadow pasture as found in the site of Rachixina (Upper Trebbia Valley) that will be discussed below. In particular, the situation under the pollarded beech trees (a) can be described as a dense and rather luxuriant lawn that prefers fresh and deep soils, whose composition exceeds the 30 species of herbaceous plants with the dominance of *Festuca*

Figure 4.
Beech trees distribution and shape in a wooded meadow-pasture (detail of the central slope at the spring of the Secchia river Valley, ca.1590 A.S. Parma, Confini, b.205/7b, dis.887).

rubra with which other good forage grasses are associated, along with proteins-rich pulses. The situation observed in (b) can be described as an impoverished stage of (a). The herbaceous layer is sometimes dishomogeneous and the number of species is lower; the physiognomic aspect is that of a meadow-pasture dominated by *Nardus stricta*, a grass rejected by cattle, and with which other bad forage species are associated, while few *Phleum alpinum* and *Festuca rubra* are still found. These positive variations induced in the herbaceous layer—certainly known by the past users of the wooded pasture—are obtained through a precise distribution of the beeches, of their number in function of the micro-topography, of the prevailing winds, of the water regime and of the fire regime controlled by shepherds. As is known, Oliver Rackham—with which the results of the observations at the site Prati di Sara were discussed—used to include the environmental and production processes that were activated in these wooded grassland in the more general study of the historical ecology of savannahs with deciduous trees; a plant typology about which there is no doubts concerning origin due to management by fire (cfr [41]). Pollen evidence from the very close site of peat bog at Prato Spilla [42, 43] has shown that the system documented in 1995 still at work at Prati di Sara became an integral part of

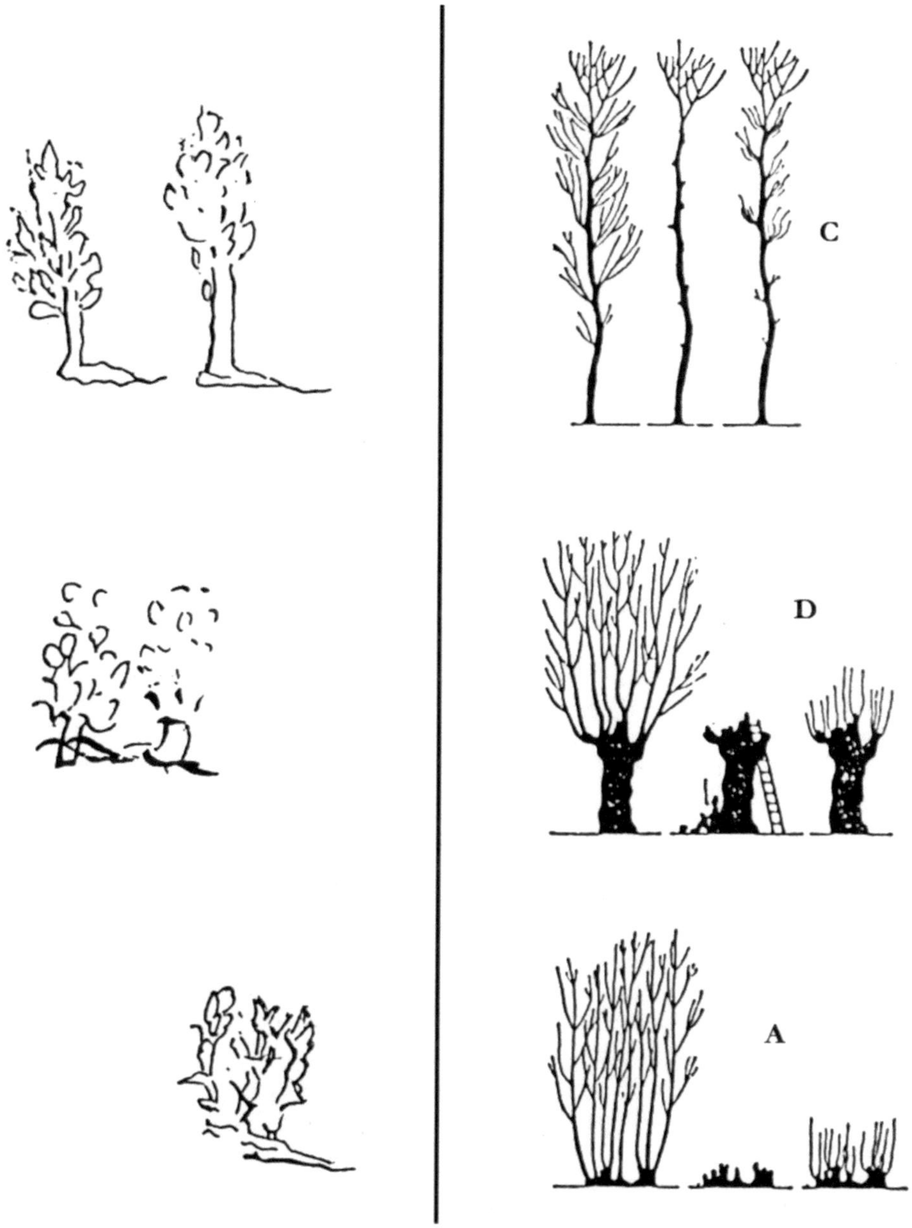

Figure 5.
A comparison between (on the left) the sixteenth century cutting practices (from the ASP 1590 map) and (on the right) the Rackham reconstruction of medieval beech wood pasture woodmanship practices: A. Stool Coppicing; D. Pollarding; C. Shredding [38].

the summer management of the higher portion of the slopes during the Lombard period. In the upper part of the pollen diagram, the basis of the M3 horizon has been radiocarbon dated to 1400 ± 45 BP, which coincides with the "Lombard Age" (sixth and seventh centuries AD) of settlements in the northern Apennines [32]. The uplands witnessed major changes in land management and settlement patterns during the Lombard phase, from an economy based upon woodland grazing (a "saltus" system, widely practiced during the Roman period) to the "alpes" beech wooded meadow system, in which livestock feeding is provided by herbs, hay, and fodder production. At the Prato Spilla site (A), the significant reduction in tree

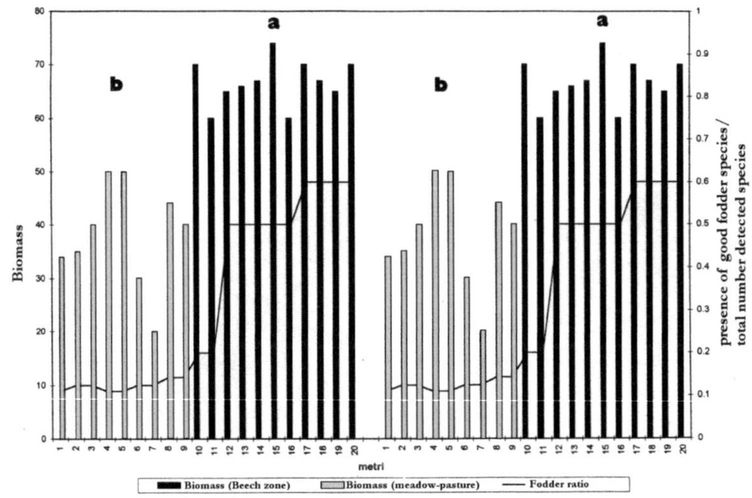

Figure 6.
Effects of the pollarded beech trees cover on the herbaceous vegetation layer of the wooded meadow at the site Prati di Sara (Upper Secchia Valley) [40].

pollen, accompanied by expansions of beech, fleaworts (*Plantago* sp.), and grass pollen reflects this change from "saltus" to "alpes" systems. Actually, the grit and sand layer at this horizon indicates that this change led to significant disruption of local soils, which in turn suggests that the established vegetation-soil associations were sensitive to change. It is unlikely that farming activities were ever intense, and they are likely to have been seasonal in the Prato Spilla area, as suggested by the temporary practice of "ronco" for the Upper Vara Valley multiple use system. It could be of interest to note that in a more recent part of the core, the pollen diagram cored at Prato Spilla shows a new expansion of Gramineae and Cyperaceae, in connection with beech pollen reduction. This possibly documents a more intensive use of the Alps slopes for hay making on bare areas as shown in the 1590 map of the individual landscape at the sources of the Secchia River and a fragment of which has been studied in the field at the site Prati di Sara.

As mentioned above, past beech wooded meadows ecology has been documented both by pollen and present floristic survey at the Rachixina site located on the W-NW side of Mount Montarlone at 334 m asl (High Trebbia valley, GE) (cfr **Figure 1**(6)). The surveyed area is part of the common land pertaining to the inhabitants of the parish of Casanova in the municipality of Rovegno. The present tree cover consists of an aged beech coppice within the Natura 2000 SIC IT 1331012 "Roccabruna." Here, a parcel of ca. 700 mq of surface presents a population of 15 pollarded beech trees: a "relict" of the wooded meadow part of an *Alpe* no longer used since the beginning of the twentieth century [40]. The parcel escaped the coppice cut for charcoal production that affected the whole common beech forest in the 1930s [44, 45] due to conflict with a supposed private property. Since then, it has not undergone other uses and the abandonment has given an unusual form to the suffering no more pollarded trees that generated a peculiar circle of radical suckers [7, 46]. The herbaceous layer underlying the beech forest is very poor due to the heavy leaf litter deposit, and it is practically absent under the pollarded trees. Anyway, some traces of past herbaceous layer can be detected observing palynological results [44]. The pollen diagram obtained in April 2007 by sampling a soil profile of about 75 cm shows the existence of three different phases attributable to variations in management of the vegetation cover [44]. The first and oldest phase (RAX 1)—that according to the radiocarbon dating should

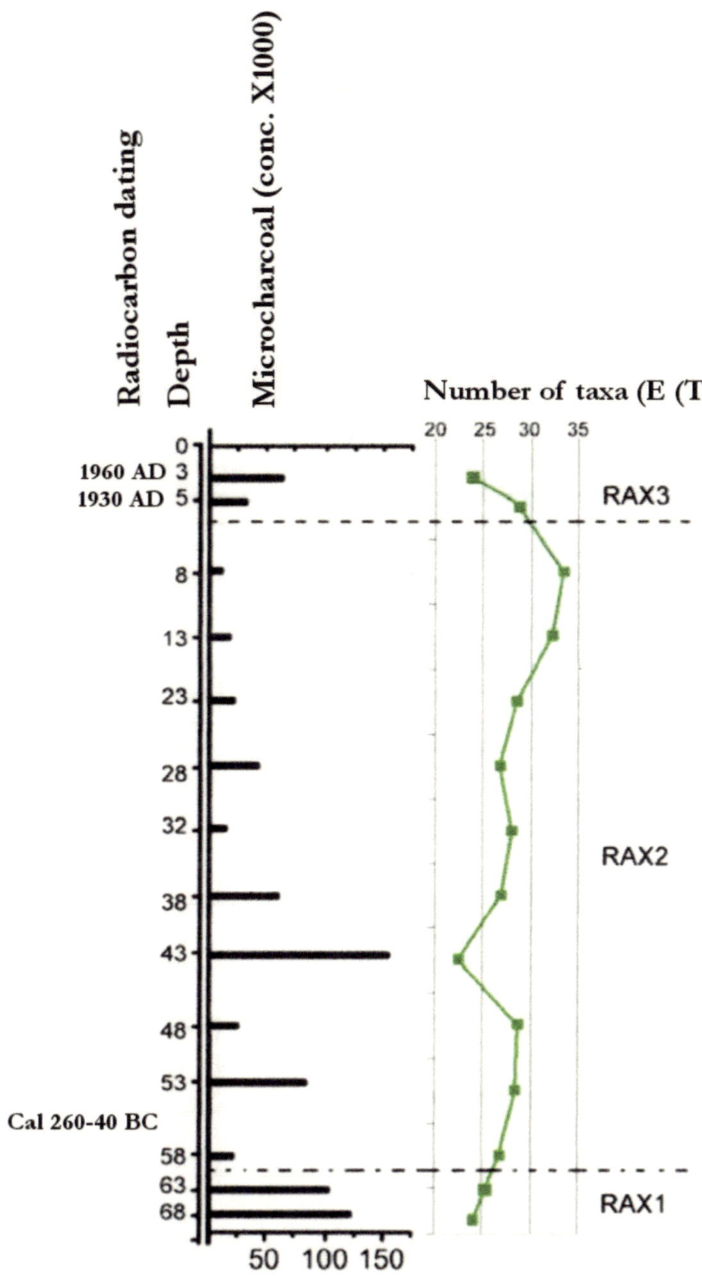

Figure 7.
The historical process of bio-diversification. The ecology of the site changes following the change in forest use: an estimate of the degree of biodiversity present in the different historical phases reconstructed by the pollen diagram at Rachixina site ([44], Figure 41).

be earlier than 280 BC—have a forest cover dominated by beech in which are present hazel, alder, deciduous oaks, pine, and silver fir. The presence of heather (*Erica* gen.) possibly proves forest grazing practices: a landscape that perfectly corresponds to the "saltus" phase documented by pollen sources at Prato Spilla site. The second phase of the pollen diagram (RAX 2) shows the dominance of beech on all tree species (silver fir practically disappears) together with the highest percentage of pollen from herbaceous species. A marker of the adoption of the wooded meadow system through a process that pollens have documented also at Prato

Spilla site, likely the *Alpe* system adopted after the Classical age. It is interesting to note that the highest degree of biodiversity of the site is recorded during the RAX 2 phase. The fire practices are shown on the columns of microcharcoal. The ecology of the site undergoes a rapid change around the first decade of 1900, when the main sheep transhumance from the coast stopped. A change that also affects the soil pH: during the centuries of use of the wood pasture, appears to be less acidic giving the N added by the flocks dropping. **Figure 7** documents the process of site bio-diversification. It shows the estimate of the degree of biodiversity present in the different phases reconstructed through the pollen diagram. The diagram, on the right side, shows the result of quantitative analysis of the microcharcoals: a peak of concentration at a depth of 43 cm could be related to the adoption of the beech wooded meadow system. Subsequently, the microcharcoals testify the pastoral controlled fire regime of the multiple system. The RAX 3 phase has been compared with archival sources and the present vegetation cover of the slope allowing to identify the living—but disappearing—heritage of this historical multiple system in the present high environmental value landscape and its problematic conservation [45–48].

4. Field and sedimentary evidences for past fir wooded pasture ecology

For the purpose of providing information for the implementation of policies concerning the future conservation through the reuse of wooded meadow-pasture systems, an ecological continuity could be established following a regressive observation from the present site ecology tracing back to the first discontinuity. It appears that in Ligurian Mountains, a main general reorganization of pastoral resources occurred in the post-classical age. In many cases, the adoption (or, on the contrary, the reduction) of specific land cover and use is precisely dated by pollen or charcoal samples (e.g., the quoted sites of Prato Spilla, Pian delle Groppere, Rachixina, Pian Brogione, Moglia di Casanova, etc.).

It is almost certain that the wooded meadow-pasture system is related to the common use of land and to the adoption of the new legal access to the collective resources that had been put in place in NW Apennines with the Lombard age customary law system. But each individual landscape of wooded meadow-pasture requires its own biographical reconstruction. On the basis of pollen and micro-charcoals evidence, the hypothesis that a wooded pasture populated with silver firs (*Abies alba*) was already in use in the Ligurian-Emilian Apennines during proto and prehistoric ages has been advanced (see [49–51]). This fact could be of interest in pointing out the problem of the ecological continuity within the present vegetation cover of these archeologically reconstructed systems.

The pollen core (**Figure 8**) at "Moglia di Casanova" has been sampled in a pond/peat bog site, located in the common land of the mentioned parish of Casanova di Rovegno on the watershed between the Trebbia and the Aveto Valleys. The collapse of *Abies* spp. percentages recorded in the pollen diagrams obtained in this watershed (Moglia di Casanova site and Mogge di Ertola site) dates at the Roman Age; while in other sites of the Aveto valley (Lago Riane and Prato Mollo), the substitution *Abies-Fagus* appears to be much older (4000 and 3000 BC, see [51–53]). The "Moglia di Casanova" peat bog site offers another significant pollen evidence: fir trees present during the Bronze Age were affected by a clearing phase which resulted in an open wood pasture rather than a closed forest [49]. Gillian Cruise, thanks to the increase in the pollen of Composite, Gramineae, Cyperaceae, and minor occurrences of Liliacee, *Melampyrum*, and *Rhinanthus*, suggested an important phase dating back to the Bronze Age of permanent pasture in (or between) the fir woods. The presence

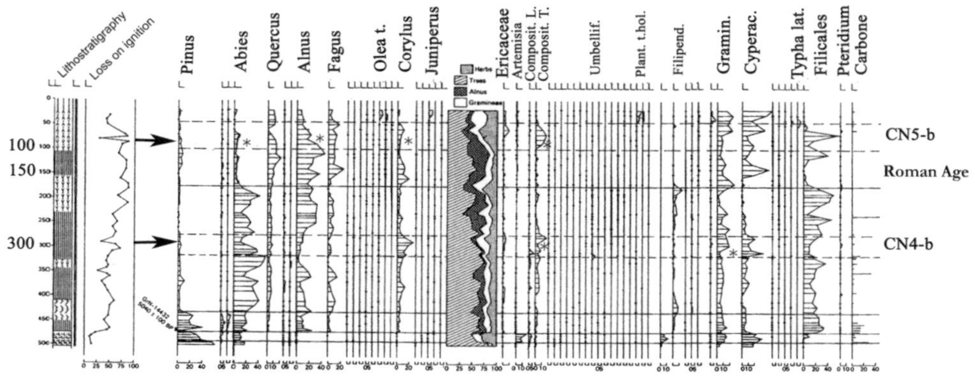

Figure 8.
Pollen diagram cored at "Moglia di Casanova" pond/peat bog site ([49], modified [7]).

of *Avena* (oats) pollen in this context had also suggested temporary small-scale fields in woodland clearings and glades, allowed by the functioning of a multipurpose system [7, 49]. Was in pre Classical Age the silver fir population managed as a multipurpose system similar to the local system documented in medieval and post-medieval ages?

In fact, at different sites of this Trebbia-Aveto watershed, silver fir continued to be present even during medieval and post-medieval ages. An interesting example comes from the site of "Pian delle Gròppere" (close to Moglia di Casanova), at 1250 m asl, where archeological documentation indicates the presence of silver fir in the arboreal vegetation during the Early Middle Ages in the context of clearance cairns for temporary fields. The abundance of juniper and fir charcoals fragments found in the basal stratigraphic unity (US) of the excavated stone mound (cairns = dial. "groppere") indicates, in fact, an episode of fire dating back to 640–770 AD; probably fire practices for agricultural purposes (early medieval practice of temporary agriculture on a wooded soil with controlled fire, called "ronco" in medieval and post-medieval texts [26, 54–57]).

Sporadic silver fir trees are present in continuity with the medieval silver fir population on the same slopes of the Upper Trebbia Valley during the nineteenth century. Textual sources help: Alberto Nota, a deputy in the administration of newly established "Provincia di Bobbio" of the Kingdom of Sardinia in 1822, quoting the trees present in the Casanova mountains, wrote: "they are seen in that surroundings also firs trees which are no longer found in the Bobbiese, although from the old buildings it is recognized that it was once a great use of such wood for construction" [7]. Large scale maps made by military staff in the first half of the XIX century are important documents to reconstruct local land use; on these slopes, the abbreviation "BP", that means "wooded pastures" ("Bosco pascolato"), was used by the topographers to indicate the importance and extension of this system. Although referred mainly as beech wooded pasture [29, 46, 58]—probably it includes sites of the existing silver fir population. Could an ecological continuity be established concerning the present silver fir population in the nearby Upper Ceno-Nure valleys? Actually, in the botanical literature, the present living populations of silver fir at the site of "Tana di Monte Nero," between the Ceno and Nure valleys, at an altitude of about 1650 m asl, are considered to be relics. It is not due to twentieth century forestry plantation policy, but rather presenting a biological continuity with the "ancient" populations, as well as with the nearby few individuals of mountain pine (*Pinus mugo* Turra) [59]. These silver fir populations located around Mount Nero had already been described by an eighteenth century surveyor

[60] during his itinerary exploring the economic resources of Piacenza's mountains. In the summer of 2005, a field visit part of the "Wetlands Archaeology Project," a regional research project on wetlands of Liguria considered as cultural heritage sites [61–64], permitted the attribution of the oldest surviving firs aged between 200 and 300 years confirming the continuity of the observed population. On M. Nero, a rare Apennine fragment of ancient pasture with silver fir is so present, in continuity with the historical populations. The interest of this area is exceptional from an historical ecology and archeological point of view. Indeed, comparisons are possible with the "fossil" fir trees conserved in many wetlands sites located in the Trebbia-Aveto Upper valleys, and in particular the site of "Mogge di Ertola." This site preserves several waterlogged trunks buried in the peat, mainly of silver fir and beech. Excavation showed that the trees fell into the peat bog approximately 4000 BP. Some of the dated tree trunks bear marks of fire and shredding practices. The trees at the "Tana di monte Nero" site present a browsing tree line (whose inferior branches still have a scorching line). Thanks to the presence at the easily accessible site of this living specimens and recently deceased specimens, it is possible to make detailed ecological observations that also analyzes the particular "candelabrum"

Figure 9.
Tana di Monte Nero, Upper Ceno Valley, Bedonia, q. About 1650 m asl: one of the living "ancient" white firs that populate a surviving strip of wooded pasture (A. Cevasco, 13.08.2005).

morphologies (**Figures 9** and **10**); those with more vegetative apexes had probably been chewed during the trees younger years.

The most important observation is represented by lateral swellings of reaction tissues due to the practice of shredding adult trees in order to collect branches for fodder (**Figure 11A** and **B**): this feature, due to the repeated cutting scarf, can be compared with those found on the Bronze Age silver fir trunks of Ertola. It is a well-known feature to the local woodmen still practicing the cut of lateral branches (dial. "skravaa"/shredding) in the turkey oak wooded pasture (the repeated cutting scarf "husks" called dial. "bucci").

Figure 10.
Tana di Monte Nero, Upper Ceno Valley, Bedonia. About 1650 m asl: bi-secular dead specimen of silver fir presenting a candlestick shape. Possibly an effect of grazing during his youth (R. Cevasco, 13.08.2005).

Figure 11.
(A) Tana di Monte Nero, Upper Ceno Valley, Bedonia, q. 1650 m asl: detail of the lateral swelling/reaction tissues ("husks") in a living silver fir (effect of shredding) (D. Moreno, 13.08.2005). (B) Reaction tissues in a sub-fossil bronze age silver fir from the Mogge di Ertola site [52, 65].

Information collected in Bergamo Alps—Central Alps Chain—would confirm the presence of "bucci"/scarf also in the grazed silver firs of Val Brembana, thanks to oral sources that experienced direct observation of the cyclical shredding of the lateral branches of the silver fir (production of fodder) and its environmental effects (lighter) on the herbaceous layer. A first survey (M. Calegari, personal communication) conducted in Schilpario (Val Brembana) has revealed, in fact, that still in the years 1950–1960, silver fir populations were subjected to "scalvo" (dial. = shredding) by shepherds. For a reference to a similar practice and similar production in the late nineteenth century Savoy Alps, see [66].

Also the cored site of Pian Brogione is part of the common land of the parish of Casanova: the pastoral resources of the slope where disputed since the end of eighteenth century by the inhabitants of the adjoining Fontanigorda parish [26]. The core section (between 74 and 174 cm) proved the abundance of silver fir trees during fourteenth to seventeenth centuries continuously living in the site from the Middle age population. The expansion of the chestnut from c. 1650 to c. 1800 appears to occur at the expense of the previous landscape classified as silver fir wooded pasture [67]. Actually, the curve of the presence of fir trees drops to zero exactly when the chestnut population appears. There are high levels of microcharcoals in this period, which are probably associated with controlled fire that reduces Ericaceae in favor of silver fir "savannah." In addition, higher levels of microcharcoals associated with clayey silt with gravel indicate that there was active erosion of the slope [55]. It is likely that the high level of charcoal and erosion correspond to episodes of cultivation in fir wooded pasture of savannah type, which appears in these slopes during the Early Middle Ages (640–770 AD) as indicated by the study of charcoal and soil stratigraphy at Pian delle Gròppere (1250 m) [7, 55, 57, 67]. A Natura 2000 project has been developed in the common land of Casanova and specifically in the site of Pian Brogione in order to reintroduce experimentally the use of controlled fire to examine the effects on local historical biodiversity and as a tool for reactivating the wood pasture system [20, 67].

5. Post-medieval beech wooded pasture: first records from charcoal hearth soils

Located in the Colline Metallifere district (Tuscany—Italy), Montieri (as well as many other settlements in the area, for example, Gerfalco, Massa Marittima, etc.) has been an important mining area for century. During the Middle Ages, mine-workings were especially concentrated on exploiting deposits of mixed sulfides of copper, lead, and silver, used for coin production. Since the 1980, several archeological investigations (for a bibliographic summary see recent papers: [68–73]) have demonstrated the important role of these activities in the development and transformation of local settlement and economic systems until the fourteenth century, when the metallurgical economy gradually collapsed. As demonstrated by recent environmental resources archeology research [22], in the post-medieval age, the wooded resources of Montieri appears to have been exploited in a multiple land use systems based on pasture that characterized this area until the nineteenth century. Investigations have been carried out on the Poggio di Montieri (the hill where the village of Montieri is located) in order to achieve a realistic insight of the past practices of woodland resources uses and of the historical and environmental dynamics which affected this area. Using a regressive and multi-proxy approach, it was possible to characterize the historical ecology of the ancient beech wooded pasture and past land management practices. It has been possible linking archeological survey, anthracological, and dendro-anthracological analyses with historical texts and cartographic information concerning charcoal hearth soils. According to textual sources, between the sixteenth and eighteenth centuries, the Poggio di Montieri was characterized by the presence of a multiple land use system based on a complex ovine, caprine, cattle and pig rearing system. These activities were strongly integrated with chestnut growing, viticulture, and agriculture (wheat, flax and hemp). Access rights to these resources were expressly regulated by time and planned (i.e., collective use, reserve, emphyteusis, etc.) by local bylaw. Texts described a complex environmental resource management system commonly used in Poggio di Montieri, named "bandita," which consisted of a temporary renting of common land that was mainly used for grazing. The pig rearing clearly had great importance in the woodland resource management: in the Montieri "bandita" system, a mixing of chestnut wood and oak trees (most of all turkey oak and English oak) is commonly defined—in sixteenth and eighteenth century documents—as "bandita di janda" namely bandita of/for acorn. Using the information provided by textual sources, woodland management practices can also be identified: for example, in the studied area collecting branches from the chestnut tree was forbidden but only during the period of the chestnut fruit harvest (starting in September until February) and so as for leaves collection or burning (in fact leaves were used for fodder). This multiple land use means that, for example, the chestnut grove could be subjected, in different periods of the year, to different practices: chestnut harvest, pruning (cutting non-interesting branches), collection of the leaves (fodder production), temporary sowing of cereals, grazing, and use of controlled fire (to encourage the re-growth of the herbaceous layer useful for grazing). These included pig pannage in oak and chestnut woods. This system was also commonly applied to beech wood, vineyards, and cultivated fields according to their respective uses. However, the "bandita" system gradually disappears between the end of eighteenth century and the beginning of nineteenth century. Indeed, in this period, general changes in the political, economic, and social patterns in Tuscany, left the way open to land privatization causing the legal end and subsequent erosion of the local collective land management system and practices [30, 74]. Signs of the last beech wooded pasture can be detected on the upper part of the Poggio di Montieri hill (between 1051 and 900 m asl): here, present vegetation stand is characterized by a high beech wood showing evidence of past coppicing and shredding (**Figure 12**).

Archeological evidence of the past beech trees management practices in the "bandita" system is concealed in the charcoal hearth soils discovered on top of the Poggio di Montieri hill (for an in-depth study of archeological and anthracological studies related to charcoal hearth soil at Poggio di Montieri see [22]). Anthracological and dendro-anthracological analysis has been conducted on C-46 site (one of the 80 charcoal hearth soil found in different areas of the Poggio di Montieri hill) located in the present beech wood at 1000 m asl (**Figure 13**).

Stratigraphic observation shows the presence of one charring phase (testified by a dark layer of 25 cm thickness) started between 1724 and 1815 AD (**Table 1**).

Figure 14 shows the percentage of wood species found in C-46, while **Figure 15** shows the results of the diameter quantification (for the method used in this research see [75–77]). Finally, **Figure 16** shows the results of a dendro-anthracological analysis intended to gain information on wood condition before carbonization (i.e., presence of fungal hyphae, anatomical deformations, etc.) [78–82].

The C-46 charcoal fragments corpus (sum of 102 fragments) is largely dominated by beech (*Fagus sylvatica*) reaching 83%, followed by Common hornbeam (*Carpinus betulus*) (8%), field maple (*Acer campestre*) (5%), juniper (*Juniperus* sp.) (3%), and wych elm (*Ulmus* sp.) (1%). Juniper has been indicated as evidence of

Figure 12.
Ancient shredded beech in the Montieri hill (V. Pescini 21.03.2017).

charcoal burning practices, as it was probably used for covering the charcoal kiln structure; this evidence fits perfectly with the hypothesis of a past savannah-type ecology for this site (according to [24]). The 36% of fragments show bark while a hyphae contamination is poorly represented. Radial cracks are scarce as well as their frequency per cm^2 indicating a probable use of dry/seasoned wood. This can be related to the greater use of good state branches (i.e., no use of rotten wood). It could also be of interest to note that the prevailing cutting season is summer (41%): the precise time characterizing the fresh fodder production (the cutting season are as follows: Spring: 16%; Summer: 41%; Autumn: 8%; and Winter: 35%). Anthracological and dendro-anthracological analyses reveal the use of beech branches for making charcoal. The C46 site provided evidence—when information are contrasted with coeval textual and historical map contents—of a wooded pasture with a scattered beech population. Probably, branches used for charcoal

Figure 13.
General view of charcoal hearth site 46 (C-46) in the beech wood (V. Pescini 21.03.2017).

Site	Acronym	Mesh Ø	Depth	Curvature	n°rings	Pith	Bark	Cutting season	Taxa	14C BP	14C AD (calibrated; 2σ)	Observation
MTPG '17	C-46 US 1	4 mm	20 - 25 cm	1	12	1	1	Winter	Fagus sylvatica	182 ± 35	1650-1697 AD (19.2%) 1724 - 1815 AD (48.2%) 1834 - 1878 AD (6.3%) 1916 -1950 AD (21.7%)	Branch Ø 0,9 cm

Table 1.
Radiocarbon results with description of the dated charcoal fragment. Dated charcoal fragment come from the bottom of the charcoal layer.

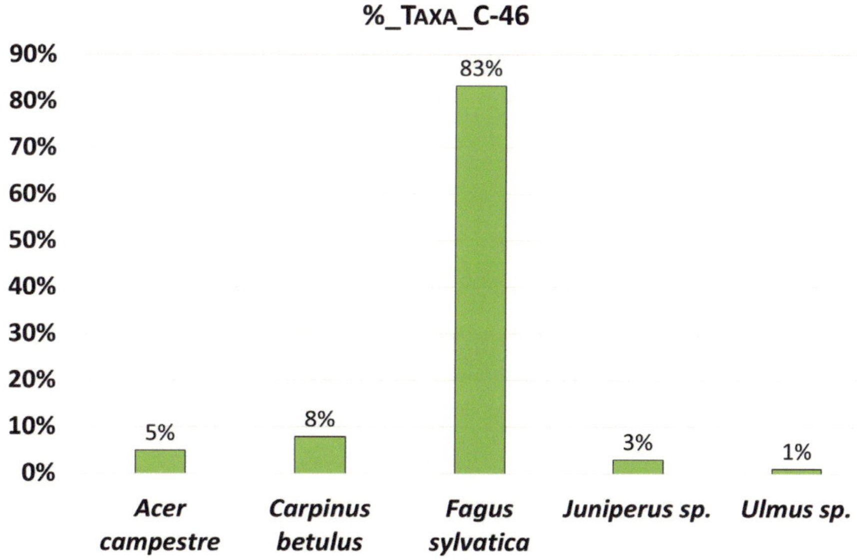

Figure 14.
Anthracological results. Graphical representation of the percentage of the species found in the C-46 charcoal hearth soil.

production derived from shredding practices: leaves were used for animal fodder or bedding and branches were burned to make charcoal. This system was documented at the Poggio di Montieri site since at least the sixteenth century, and the consequences of an intensive pastoral land use must be considered.

A similar beech wood management system has also been detected at Urbiarrate, in the Basque Country in Spain. This research is included in a wider

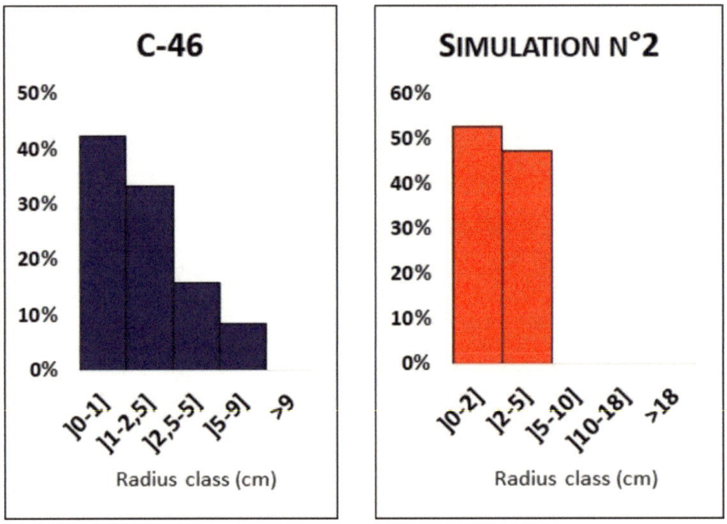

Figure 15.
Results of the wood diameter studies attesting a dominance of small wood diameters.

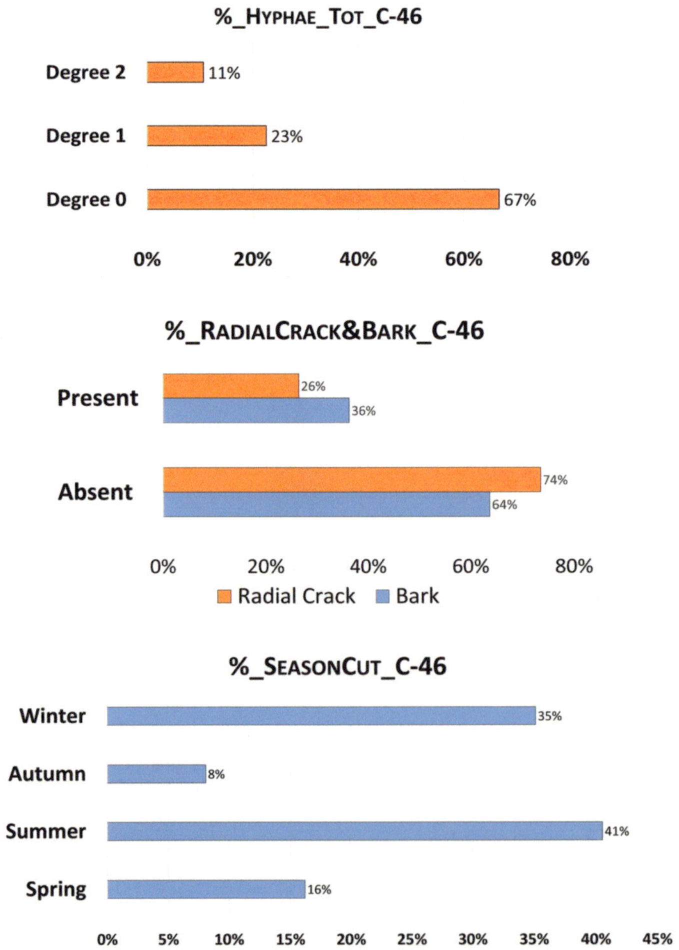

Figure 16.
Dendro-anthracological results. For hyphae results: degree 0 = missing feature; degree 1 = rare feature; and degree 2 = common feature.

Marie Curie project (2015–2016): "ARCHIMEDE—Archaeology of Commons: Cultural Heritage and Material Evidence of a Disappearing Europe." It involves different laboratories such as the Research Group of Cultural Heritage and Landscapes (GIPYPAC—University of the Basque Country), the Laboratory of Environmental Archeology and History (LASA—University of Genoa), and the FRAMESPA Terrae Laboratory (University of Toulouse Le Mirail). This project aimed to connect structures and forms of the possessory actions and the rights of use on public lands known through multiple sources (i.e., toponymy, text, cartography, etc.) with archeological evidence, including those recognized by historical ecology and sedimentary sources (palynology, anthracology, etc.) [83, 84]. One of the case studies of this project was Urbiarrate (UTM 30 N ETRS89—Lat. 42.9640769; Long. 2.3690675) in the municipality of Oñati in the Gipuzkoa province. Currently, this area is mainly occupied by pastures (used for sheep, cattle and equine grazing), mountain meadows, moors, and acidophilous beeches managed as high-standard woods. Some areas have been subjected to repopulation of coniferous species, i.e., Austrian pine (*Pinus nigra* Arnold), Douglas fir (*Pseudotsuga menziesii* Mirb.), and larches (*Larix* spp.). This area was subject in the past to a land management system providing a collective use of environmental resources, but at the same time, forms of temporary appropriation by private social actors, through the establishment/exercise of the so-called "seles." Seles are defined and circumscribed spaces inserted within the common lands and linked to specific uses that were, at least in the post-medieval period, the subject of various conflicts and which today are mostly privatized ([85, 86], in [83]). Archeological investigations were carried out inside the beech forest, in an area of about 1.4 hectares and included between 1200 and 1000 m asl, highlighting the presence of a dozens of charcoal hearth soils of different size (minimum 3–4 m of diameter and maximum 6–7 m). Two charcoal hearth soils (C-1700 and C-7000) were subjected to stratigraphic and anthracological analyses, revealing four charcoal layers identified and interpreted as different charring phases (the succession between dark layer and red/yellow clay layer show phase of production and subsequent abandonment) (**Figures 17** and **18**).

Two radiocarbon dating were made, one for each charcoal hearth soil (see **Figure 18** for the stratigraphic sequence and location of the dated charcoal fragments) in order to have a chronological range that indicates the first carbonization activities and the last ones (**Table 2**).

Figure 17.
General view of charcoal hearth soil C-7000 (V. Pescini 21.09.2016).

Figure 18.
Charcoal heart soil section. On the left: C-7000. On the right: C-1700. Red point indicates the dated charcoal fragment (V. Pescini 21.09.2016).

Site	Acronym	Taxa	Depth	14C BP	14C AD (calibrated; 2σ)
Urbiarrate	UR15_1	*Fagus sylvatica* (branches)	10/15 cm	150 ± 35	1665–1785 AD (46.6%) 1795–1895 AD (31.5%) 1905–... AD (17.5%)
Urbiarrate	UR15_2	*Fagus sylvatica* (branches)	60/70 cm	275 ± 50	1465–1680 AD (84.7%) 1760–1805 AD (8.3%) 1935–... AD (2.4%)

Table 2.
Radiocarbon results showing different phases of charcoal production from the post-medieval period up to the nineteenth century.

The anthracological spectrum shows a very precise image in both charcoal hearth site: 90% of the analyzed fragments correspond to beech, while the remaining 10% is represented by indeterminable fragments (too vitrified, deformed, etc.). Diameter measurement made on the most recent charcoal layer (US 7001) reveals the use of small and medium size log (**Figure 19**). **Figure 20** shows dendro-anthracological results made on US 7001.

Only 5% of the charcoal has bark; wood cut was made during winter. 58% of the total charcoal does not have radial fracture: dried wood seems to be the most used (testified also by the absent of cellular collapse). 52% of the total charcoal has only few hyphae, so wood was scarcely attacked by mushrooms. Thanks to these information, it is possible to hypothesize the use of beech pruning wood (i.e., pollarding and shredding) in a multiple beech wood management system, which provides the contemporary presence of grazing and charcoal production. An evidence of this practice is still visible in the bearing of some old beech specimens (**Figure 21**).

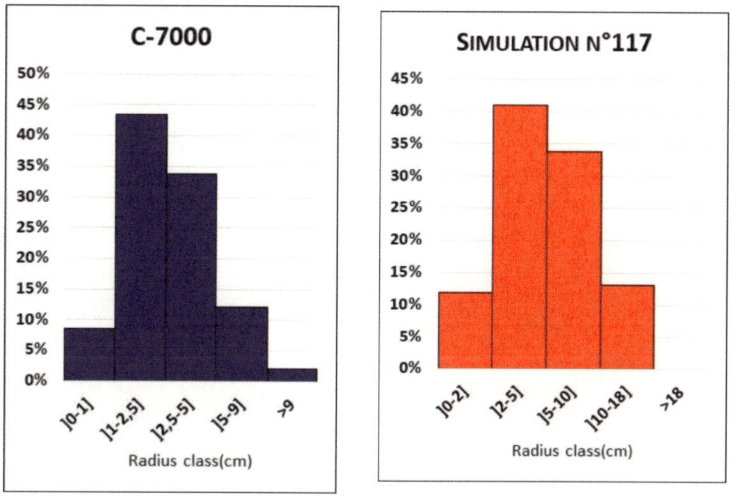

Figure 19.
Results of the wood diameter studies attesting a dominance of small wood diameters.

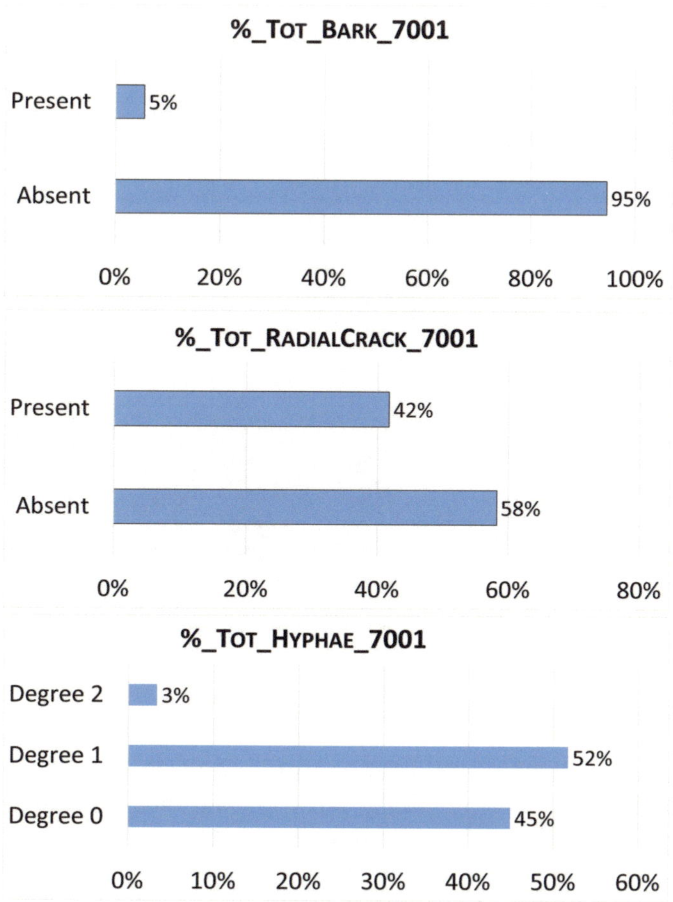

Figure 20.
Dendro-anthracological results. Percentages are calculated on a sum of 91 charcoal fragments. For hyphae results: degree 0 = missing feature; degree 1 = rare feature; and degree 2 = common feature.

Figure 21.
Ancient shredded beech at Urbiarrate Gipuzkoa province (A. Stagno 21.09.2016).

6. Conclusions

There is an urgent need to address the problem of Mediterranean woodlands: in the last decades, the idea of rewilding and afforestation in rural areas to replace abandoned agricultural lands has been put into practice in several parts of Europe despite some criticisms. Ecologists, such as [87, 88], discussed the idea of a pristine "natural forest" across the continent, suggesting the presence of a shifting mosaic of open areas, patches of mature trees, and patches of regeneration. Moreover, as a consequence of rural depopulation and land abandonment, Mediterranean and Alpine areas have been subjected to a rapid spread of spontaneous woodlands. While some scholars value this process as a welcome return to a past before the agricultural and pastoral human activities started to manage the land, others consider that the new woodland disguises the loss of traditional woodland knowledge and the reduction of biodiversity associated with pastures and meadows [20, 89]. In the case studies presented above, the medieval and post-medieval history of the wooded resources of the Mediterranean mountains appears to be dominated by animal production; a complex agro-sylvo-pastoral system concealed by the development of nineteenth and twentieth century forestry sciences and allied forestry law regime that bring the end to the customary regime, which was used to regulate woodland and commons resources access.

Thanks to the research presented in this chapter, some observations can be made.

1. If we abandon the models of interpretation of the structural ecology—where the reference is to an untouched, pristine ecosystem—and adopt the perspective of historical ecology, an unsuspected historical dynamic of transformations processes (intensifications/abandons, bio-diversification, etc.) that have affected Mediterranean mountain woodland resources systems emerges. As a consequence, the importance of past economics relating to a number of different local farming/breeding systems can be revealed, applied in their management as is the case of wooded pasture-meadows systems. This evidence can have a great significance in the identification and managing of many aspects of the present landscape and resource heritage. For example, for the rural and the gastronomic heritage of an area, discarding the ambiguous concept of a bio-cultural heritage.

2. The recent development of the Environmental Resource Archaeology (ERA)—an environmental archeology that includes the historical ecology regressive approach—requires precise historical models for the understanding of the past functioning of resources system. Conversely, it offers documentation that enables the geographer to identify the historical practices of production and activation coupled with the management system of pastoral resources. Generally, a new historical understanding of the evidence produced by field research is possible by assuming ERA procedures. On the other hand, the contrast between sedimentary sources (field sources) (palynology, anthracology, archeozoology, etc.) allows a "realistic deciphering" of the content of the texts themselves.

3. Comparing sources is apparently an obvious methodological achievement of multidisciplinary research, but it is exclusively possible through the adoption of a local, topographical, observation scale both for the environmental and for the social context network as suggested by the microhistorical approach. Unfortunately, this is a choice that does not seem desirable in the fields of cultural geography and history [16]. Cultural history (and geography or historical-cultural geography) offered for years effortless—but highly misleading—generalizations based on a rather symbolic or metaphorical "cultural decipherment" of evidence and from the perspective of a global scale history. Cultural history and geography have a real difficulty in coping with and assimilating developments in historical ecology [90].

4. Local systems for the production and activation of the Apennine wooded resources are (agro-sylvo-pastoral) multiple systems. During the customary regime—until the adoption of the Forestry Codes and Civil Codes of laws in the mid-nineteenth century—fodder production was therefore ubiquitous in

agricultural and wooded use of land. Specialized forms operating with particular production/control practices have existed to achieve an intensification and improvement of the mere fodder production of adding hay, grass, and fruits for human and animal nourishment, as is the case of wooded meadows and pasture discussed in this chapter.

Acknowledgements

We thank R. Hearn and C. Montanari for comments on the manuscript. The authors wants to thank the staff of the State Archive of Parma for their help in archive research.

Author contributions

While the study has been started by D.M., this paper is the result of common work by all authors: R.C the Sections 2 and 4, N.C the Sections 1 and 6, and V.P. the Sections 3 and 5.

Author details

Diego Moreno[1], Roberta Cevasco[2,3], Valentina Pescini[1*] and Nicola Gabellieri[3,4]

1 DAFIST/DISTAV, Laboratory of Archaeology and Environmental History, University of Genoa, Italy

2 University of Gastronomic Science, Pollenzo, Italy

3 Laboratory of Archaeology and Environmental History, University of Genoa, Italy

4 University of Trento, Italy

*Address all correspondence to: valpes87@gmail.com

References

[1] Rackham O. Woodmanship. Botanical Journal of Scotland. 1994;**46**(4):651-657. DOI: 10.1080/13594869409441777

[2] Rackham O, Moody J. The Making of the Cretan Landscape. Manchester: Manchester University Press; 1996. 237 p

[3] Bell M, Limbrey S. Archeological aspects of woodland ecology. Journal of Archaeological Science. 1982;**10**(3):293-294

[4] Richer S, Gearey B. From Rackham to REVEALS: Reflections on palaeoecological approaches to woodland and trees. Environmental Archaeology. 2018;**23**(3):286-297. DOI: 10.1080/14614103.2017.1283765

[5] Reitz E, Shackley M. Environmental Archaeology. Manuals in Archaeological Method, Theory and Technique. London, New York: Springer; 2012. 516 p

[6] Rackham O. The History of the Countryside. London: J.M. Dent & Sons; 1986. 448 p

[7] Cevasco R. Memoria verde, nuovi spazi per la geografia. Diabasis: Reggio Emilia; 2007. 320 p

[8] Raggio O. Microhistorical approaches to the History of Liguria; from Microanalysis to Local History. Edoardo Grendi's Achievements. In: Balzaretti R, Pearce M, Watkins C, editors. Ligurian Landscape. Studies in Archaeology, Geography & History in Memory of Edoardo Grendi. London: Accordia Research Institute, University of London; 2004. pp. 97-104

[9] Ingold A. Écrire la nature. De l' histoire sociale à la question environnementale? Annales Histoire Sciences Sociales. 2011;**66**:11-29

[10] Moreno D. Entre géographie, écologie et histoire : Pour une micro-analyse géographico-historique. Actes Colloque Géohistoire de l'environnement et des paysages, Toulouse. 2016 (in press)

[11] Cevasco R, Gabellieri N, Pescini V. Une approche historique et archéologique pour l'étude des systèmes de gestion des ressources environnementales: Expériences en Ligurie (Italie N-O). Actes Colloque Géohistoire de l'environnement et des paysages, Toulouse. 2016 (in press)

[12] Szabò P. Why history matters in ecology: An interdisciplinary perspective. Environmental Conservation. 2010;**37**:380-387

[13] Rotherham ID, editor. Cultural Severance and the Environment. The Ending of Traditional and Customary Practice on Commons and Landscapes Managed in Common. Dordrecht: Springer; 2013. 447 p. DOI: 10.1007/978-94-007-6159-9

[14] Cevasco R, Moreno D, Hearn R. Biodiversification as an historical process: An appeal for the application of historical ecology to bio-cultural diversity research. Biodiversity and Conservation Journal. 2015;**24**(13): 3167-3183. DOI: 10.1007/s10531-015-0943-3

[15] Hoskins WG. The Making of the English Landscapes. London: Hodder and Stoughton; 1955. 240 p

[16] Torre A. Spatial turn in history. Paysages, regards, ressources pour une historiographie de l'espace. Annales Histoire Sciences Sociales. 2008;**63**:1127-1144

[17] Torre A. Places. Production of locality in early modern and modern society. London: Routledge; (in press)

[18] Grove AT, Rackham O. The Nature of Mediterranean Europe: An Ecological History (Second Printing with Corrections Ed.). New Haven and London: Yale University Press; 2003. 384 p

[19] UNESCO-SCBD. Florence Declaration on the Links Between Biological and Cultural Diversity, Unesco—Scbd Joint Program Between Biological and Cultural Diversity. 2014. Available from: https://www.reterurale.it/flex/cm/pages/ServeBLOB.php/L/IT/IDPagina/14388

[20] Kirby K, Watkins C, editors. The Ecological History of European Forests. Wallingford: CAB International; 1998. 384 p

[21] Parrotta J, Agnoletti M, Johann E, editors. Cultural heritage and sustainable forest management: The role of traditional knowledge. In: Proceedings of the IUFRO Conference held in Florence, Italy, 8-11 June 2006. Warsaw: MCPFE Liaison Unit; 2006

[22] Pescini V. Archeologia delle risorse ambientali: Le fonti antracologiche per la storia del paesaggio e dell'ambiente [thesis]. Università di Genova; 2018

[23] Agnoletti M. Environmental thinking and cultural values: A reflection on environmental globalization and the Mediterranean culture. Global Environment. 2014;**7**:13-46

[24] Rackham O. Prospects for landscape history and historical ecology. Landscape. 2000;**1-2**:3-17. DOI: 10.1179/lan.2000.1.2.3

[25] Rackham O. Woodlands. New Naturalist Series. London: Harper Collins; 2006. 592 p

[26] Beltrametti G, Cevasco R, Moreno D, Stagno AM. Les cultures temporaires entre longue durée et chronologie fine (montagne ligure, Italie). In: Viader R, Rendu C, editors. 34 Journées internationales d'histoire. Cultures temporaires et féodalité. Les cycles culturaux et l'appropriation du sol dans l'Europe médiévale et moderne, Abbaye de Flaran; Presses Universitaires du Mirail. 2014

[27] Bertolotto S, Cevasco R. The "Alnocoltura" system in the Ligurian Eastern Apennines: Archive evidence. In: Agnoletti M, Anderson S, editors. Methods and Approaches in Forest History. Vol. 3. Wallingford: Cabi Publishing; 2000. pp. 189-202

[28] Moreno D. Dal documento al terreno. 2nd ed. Genova: Genova University Press; 2018. 456 p

[29] Moreno D. Une source pour l'histoire et l'archéologie des ressources végétales. Les cartes topographiques de la montagne ligure (Italie). In: Bousquet-Bressolier C, editors. L'œil du cartographe et la représentation géographique du Moyen Age à nos jours. Vol. 18. Paris: CTHS; 1995. pp. 175-197

[30] Parigino GV, editor. Beni comuni e strutture della proprietà. Dinamiche e conflitti in area toscana tra basso medioevo e età contemporanea. Firenze: Associazione di Studi Storici Elio Conti; 2017

[31] Gabellieri N., Tinterri D. The mathematician and the beans. Food productions, hydrogeological system and the struggle for water in the XVIII century Val Bisagno (Genova, Italy). In: Cevasco R, Hearn R, editors. Local Foodways (in press)

[32] Davite C, Moreno D. Des saltus aux alpes dans les Appennins du Nord (Italie). Une hypothèse sur la phase du Haut Moyen Age (560-680 ap. J.C.) dans le diagramme pollinique du site de Prato

Spilla. In: Colarelle M, editor. L'homme et la Nature au Moyen Age. Paris: CNRS; 1996. pp. 138-142

[33] Raggio O. Feuds and State Formation, 1550-1700 the Backcountry of the Republic of Genoa. London: Palgrave McMillan; 2018. 317 p

[34] Moreno D, Cevasco R, Bertolotto S, Poggi G. Historical ecology and post-medieval management practices in alder woods (*Alnus incana* (L.) Moench) in the northern Apennines, Italy. In: Kirby K, Watkins C, editors. The Ecological History of European Forests. Wallingford: CAB International; 1998. pp. 185-201

[35] Cevasco R. The environmental heritage of a past cultural landscape: The Alderwoods (*Alnus incana* Moench) in the upper Aveto Valley (NW Apennines). In: Armiero M, Hall M, editors. Nature and History in Modern Italy. Ohio: Ohio University Press; 2010. pp. 126-140

[36] Molinari C, Montanari C. Interdisciplinary approach for reconstructing an alder-based historical agricultural practice of the Eastern Ligurian Apennines (NW Italy). Environmental Archaeology. 2015;**21**(1):31-44. DOI: 10.1179/1749631414Y.0000000056

[37] Stagno AM. Lo spazio locale dell'archeologia rurale. Risorse ambientali e insediamenti nell'Appennino ligure tra XV e XXI secolo. Firenze: All'Insegna del Giglio; 2018

[38] Rackham O. Boschi e storia dei sistemi silvo-pastorali in Inghilterra. Quaderni storici. 1982;**49**:16-48

[39] Agnoletti M. Boschi da pascolo e pascoli arborati. In: Storie del bosco. Il paesaggio forestale italiano. Bari: Laterza; 2018. pp 44-100

[40] Moreno D, Poggi G. Storia delle risorse boschive nelle montagne mediterranee: Modelli di interpretazione per le produzioni foraggere in regime consuetudinario, Atti della XXVII settimana di Studi "L'uomo e la foresta sec. XIII-XVIII". Prato: Le Monnier; 1996. pp. 635-653

[41] Rackham O. Forest history of countries without much forest: Question of conservation and Savanna. In: Cavaciocchi S, editor. L'Uomo e la foresta, secc XIII-XVIII. Firenze: LeMonnier; 1995. pp. 297-326

[42] Lowe JJ, Davite C, Moreno D, Maggi R. Holocene pollen stratigraphy and human interference in the woodlands of the northern Apennines, Italy. The Holocene. 1994;**4**:153-164

[43] Lowe JJ, Davite C, Moreno D, Maggi R. Stratigrafia pollinica olocenica e storia delle risorse boschive dell'appennino settentrionale. Rivista Geografica Italiana. 1995;**102**:265-308

[44] Molinari C. Ricerche palinologiche per l'identificazione di sistemi agro-silvo-pastorali storici [thesis]. Doctoral School in Historical Geography for the Enencement of Environmental Heritage, University of Genoa; 2010

[45] Cevasco R, Molinari C. Microanalysis in woodland historical ecology. Evidences of past leaf fodder production in NW Apennines (Italy). In: Saratsi E, Bürgi M, Johann E, Kirby KJ, Moreno D, Watkins C, editors. Woodland Cultures in Time and Space. Athens: Embryo Publications; 2009. pp. 147-154

[46] Cevasco R. La copertura vegetale dell'alta val Trebbia nelle ricognizioni topografiche del Corpo di Stato Maggiore Sardo (1816-1852). Approccio storico all'ecologia dei siti. Archeologia Postmedievale. 2002;**6**:195-214

[47] Cevasco R., Moreno D. Historical ecology in modern conservation in Italy. In: Kirby K, Watkins C. editors. Europe's Changing Woods and Forests: From Wildwood to Managed Landscapes, CAB International; 2015. pp. 227-242

[48] Cevasco R, Moreno D. Wooded pastures hay meadows in the Ligurian-Tuscan-Aemilian Apennines, Italy. In: Krzywinski K, O'Connell M, Küster HJ, editors. Cultural Landscapes of Europe, Fields of Demeter, Haunts of Pan. Bremen: Aschenbeck Media; 2009. pp. 104-105

[49] Cruise GM. Environmental change and human impact in the upper mountain zone of the Ligurian Apennines: The last 5000 years. Rivista di Studi Liguri. 1991;**LVII**(1-4):175-194

[50] Branch NP. Late Würm Lateglacial and Holocene environmental history of the Ligurian Apennines, Italy. In: Balzaretti R, Watkins C, Pearce M, editors. Ligurian Landscapes, Studies in Archeology, Geography & History. London: Accordia Research Institute, University of London; 2004. pp. 7-69

[51] Maggi R. I monti sun eggi: The making of the Ligurian landscape in prehistory. In: Balzaretti R, Watkins C, Pearce M, editors. Ligurian Landscapes: Studies in Archaeology, Geography & History. London: Accordia Research Institute; 2004;**10**:71-82

[52] Maggi R. I monti sono vecchi. Archeologia del paesaggio dal Turchino alla Magra. Genoa: De Ferrari; 2016. 223 p

[53] Branch NP. Early-Middle Holocene vegetation history, climate change and human activities at Lago Riane (Ligurian Apennines, NW Italy). Vegetation History and Archaeobotany. 2013;**22**:315-334

[54] Cevasco R. Multiple use of tree-land in the Northern Apennines during post medieval times. Linking clues of evidence. In: Balzaretti R, Watkins C, Pearce M, editors. Ligurian Landscapes: Studies in Archaeology, Geography & History. London: Accordia Research Institute, University of London; 2004. pp. 155-177

[55] Branch N, Guido MA, Menozzi BI, Placereani S, Montanari C. Prime analisi polliniche per il sito «Moggia di Pian Brogione» (Casanova di Rovegno, Ge). Archeologia Postmedievale. 2002;**6**:125-131

[56] Campana N, Cevasco R. Un'area di interesse archeologico ambientale: Studi di ecologia storica a "Pian delle Gròppere" (Casanova di Rovegno, Genova). In: Stringa P, Moreno D, editors. Patrimonio storico-ambientale. Esperienze, progetti e prospettive per la valorizzazione delle aree rurali. Rapallo: Officine Grafiche Canessa; 2001. pp. 225-232

[57] Balzaretti R. Dark Age Liguria. London: Bloomsbury; 2013. 192 p

[58] Marullo E. Risorse pastorali dell'alta val Trebbia nella cartografia topografica alla metà del XIX secolo. Archeologia Postmedievale. 2002;**6**:103-110

[59] Alessandini A, De Marchi A, editors. Il Monte Nero. Bolonia: Regione Emilia-Romagna; 1984. 107 p

[60] Boccia A. Viaggio ai monti di Parma, 1804. In: Capacchi G, editor. Parma: Palatina Editrice; 1989

[61] Cevasco A, De Pascale A, Guido MA, Montanari C, Maggi R, Nicosia C. Le Mogge di Ertola (Appennino ligure): Un contributo all'archeologia del fuoco e all'archeologia dell'acqua. In: Cevasco R, editor. La natura della montagna. Sestri Levante: Oltre; 2013. pp. 413-427

[62] Guido M-A, Menozzi B-I, Bellini C, Placereani S, Montanari C. A

palynological contribution to the environmental archaeology of a Mediterranean mountain wetland (North West Apennines, Italy). The Holocene. 2013;**23**:1517-1527

[63] Bellini C, Cevasco R, Moreno D, Guido MA, Montanari C. The Mogge di Ertola Site, High Aveto Valley, Ligurian Apennines (N. Italy): Bronze age, medieval and present cultural landscapes. In: Krzywinski K, O'connell M, Küster HJ, editors. Cultural Landscapes of Europe. Fields of Demeter, Haunts of Pan. Bremen: Aschenbeck Media; 2009. pp. 108-109

[64] Menozzi BI, Bellini C, Cevasco A, Cevasco R, De Pascale A, Guido MA, et al. The archaeology of a peat bog in context: Contribution to the study of biodiversification processes in historical time (Ligurian Apennine, NW Italy). Proceedings of the 4th International Congress of Medieval and Modern Archaeology. Paris: Medieval Europe; 2007. Available from: http://medieval-europe-paris-2007.univ-paris1.fr

[65] Cevasco R Dopo Sereni: Dal paesaggio agrario al patrimonio rurale. Le nuove fonti. In: Quaini M, editor. Paesaggi agrari. L'irrinunciabile eredità scientifica di Emilio Sereni. Cinisello Balsamo: Silvana Editoriale Spa; 2011. pp. 161-170

[66] Briot F. Nouvelles études sur l'Economie alpestre. Paris; 1907

[67] Cevasco R, Moreno D, Balzaretti R, Watkins C. Historical chestnut cultures, climate and rural landscapes in the Apennines. In: Harvey D, Perry J, editors. The Future of Heritage as Climates Change. Vol. 8. London: Routledge; 2015. pp. 130-147

[68] Benvenuti M, Bianchi G, Bruttini J, Buonincontri M, Chiarantini I, Dallai L, et al. Lo studio del comprensorio minerario delle Colline metallifere toscane: Un approccio interdisciplinare. In: 9th International Symposium on Archaeological Mining History. Research and Preservation of the Ancient Mining Areas, 5-8 giugno Trento. 2014. pp. 261-287

[69] Bianchi G. Archeologia della signoria di castello (X-XIII secolo). In: Gelichi S, editor. Quarant'anni di archeologia medievale in Italia. La rivista, i temi, la teoria, i metodi, Firenze: All' Insegna del Giglio; 2014. pp. 157-172

[70] Bianchi G. Analysing fragmentation in the early Middle Ages: The Tuscan model and the countryside in North-Central Italy. In: Gelichi S, Hodges R, editors. New Directions in Early Medieval European Archaeology. Essays for Riccardo Francovich, Turnhout. Oxford: Archaeopress Publishing Ltd.; 2014. pp. 301-335

[71] Bianchi G. Recenti ricerche nelle Colline Metallifere ed alcune riflessioni sul modello toscano. Archeologia Medievale. 2015;**42**:9-26

[72] Bruttini J. Minatori e signorie nelle Colline Metallifere [thesis]. Università di Siena; 2013

[73] Dallai L. Archeologia delle attività produttive e metallurgiche. Il caso toscano: Le colline metallifere grossetane. In: Porras AG, editor. Arqueología de la producción en épocamedieval, Granada: Alhulia; 2013. pp. 291-304

[74] Tocchini L. Usi civici e beni comunali nelle riforme leopoldine. Studi Storici. 1961;**2**(II):223-266

[75] Paradis-Grenouillet S, Dufraisse A. Radius curvature measurements and wood diameter: A comparison of different image analysis techniques. In: Damblon F, editor. Proceedings of the Fourth International Meeting of

Anthracology: Brussels, 8-13 September 2008. Royal Belgian Institute of Natural Sciences; 2013

[76] Paradis-Grenouillet S, Alleé P, Servera Vives G, Ploquin A. Sustainable management of metallurgical forest on Mont Lozère (France) during the Early Middle Ages. Environmental Archaeology. 2015;**20**:168-183. DOI: 10.1179/1749631414Y.0000000050

[77] Paradis Grenouillet S. Etudier le "foretes metallurgiques": Analyses dendroanthracologiques et approaches geohistoriques. Exemple des forêts du mont Lozere et du Perigord Limousin, [thesis]. Universite de Limoges; 2012

[78] Py V, Durand A, Ancel B. Anthracological analysis of wood fuel used for firesetting in medieval metallic mines of the Faravel district (southern French Alps). Journal of Archaeological Science. 2013;**40**:3878-3889

[79] Moskal-del Hoyo M, Wachowiak M, Blanchette RA. Preservation of fungi in archaeological charcoal. Journal of Archaeological Science. 2016;**37**:2106-2116

[80] Fouedjeu Foumou L. Histoire des forêts et du charbonnage dans les Pyrénées ariégeoises: Analyses anthracologique et dendro-anthracologique des plateformes de charbonnage de la forêt de Bernadouze (Suc-et-Sentenac), Mémoire de Master 2, Università Paul Valery Montpellier. 2015-2016

[81] Théry-Parisot I. Economie des combustibles au Paléolithique: Expérimentation, taphonomie, anthracologie. Paris: CNRS; 2001

[82] Marguerie D, Hunot J-Y. Charcoal analysis and dendrology: Data from archaeological sites in North-Western France. Journal of Archaeological Science. 2007;**34**:1417-1433

[83] Stagno AM. Archaeology of Commons: A multidisciplinary approach to the reconstruction of multiple uses and conflicts on European uplands. In: Third International Landscape Archaeology Conference 2014 Proceedings. 2016. DOI: 10.5463/lac.2014.21>

[84] Stagno AM. Montes de Utilidad Pública de Álava y Gipuzkoa: Sierra de Aizkorri (Parzonería General de Gipuzkoa y Álava, Oñati), comunales de Zalduondo, montes comunes de Salvatierra, Narbaiza (San Millán), Heredia (Barrundia) y Zuazo de San Millán (San Millán), Arkeoikuska 2015. 2016. pp. 394-405

[85] Díaz de Durana JR, Fernández De Larrea JA. Economía ganadera y medio ambiente. Guipúzcoa y el Noreste de Navarra en la Baja Edad Media. Historia Agraria. 2002;**27**:43-64

[86] Díaz De Durana JR. Para una historia del monte y del bosque en la Guipuzcoa bajomedieval: Los seles. Titularidad formas de cesión y de explotación. Anuario de Estudios Medievales. 2001;**31**(1):49-73

[87] Rackham O. Implications of historical ecology for conservation. In: Sutherland WJ, editor. Conservation Science and Action. Malden: Blackwell Science; 1998. pp. 152-175

[88] Vera F. Grazing Ecology and Forest History. Wallingford: CABI; 2000. 527 p

[89] Cevasco R, Moreno D. Rural landscapes: The historical roots of biodiversity. In: Agnoletti M, editor. Italian Historical Rural Landscapes: Cultural Values for the Environment

and Rural Development. Dordrecht: Springer; 2013. pp. 141-152

[90] Eriksson O. What is biological cultural heritage and why should we care about it? An example from Swedish rural landscapes and forests. Nature Conservation. 2018;**28**:1-32. DOI: 10.3897/natureconservation.28.25067

Chapter 3

Pollard Forest of *Fraxinus angustifolia* in the Centre of Iberian Peninsula: Protection and Management

Fernando Allende Álvarez, Gillian Gomez-Mediavilla and Nieves López-Estébanez

Abstract

The activity of pollarding *Fraxinus angustifolia* L. is one of the most singular in the Guadarrama piedmont, Madrid. This treatment generates a patrimonial landscape of great interest and identity. The main aims of this work are to determine the surface covered by *Fraxinus angustifolia* L.; its conservation state and structure; the main variables that influence its distribution; and the elaboration of a typology. We start with the historical construction of these forests (from the eleventh century until the present moment). Some of them are excellent examples of pollard forests, thanks to the traditional regulations based on regional codes and byelaws and on a rational exploitation from some owners. Then, we analyse the physical determinants of this forest distribution. We establish a typology in which we consider the surface soil moisture, accompanying species, slope, morphostructural unit and type of property. *Fraxinus angustifolia* L. is, due to its narrow dependence of the existence of water, an indicator species of climate change and therefore vulnerable to the increase in temperatures and its consequences. Furthermore, its value as cultural heritage and the need for its conservation for its silvicultural values are highlighted.

Keywords: *Fraxinus angustifolia* L., silvopastoral system, pollard, cultural heritage, surface soil moisture

1. Introduction

Pruning and pollarding of tree species constitute an activity that has existed in European forests for centuries. Pruning and pollarding of tree species constitute an activity that has existed in European forests for centuries. Although pruning is still a frequent silvicultural practice, in some areas, some species are currently no longer pruned and there has been a decline in other uses that had been implemented in forests, which played a vital role in shaping their silvo-structures [1–3]. This pruning responded to different objectives, among these, production of wooden beams, charcoal, firewood, timber for ships, leaves for fodder, etc. All these activities were tightly controlled by traditional regulations [4]. These coppice forests present an "aerial" morphology [5]. In general terms, among the formations in which these practices were abandoned, we can highlight forests of oak (*Quercus pyrenaica* Wild., *Quercus robur* L.), birch (*Betula* ssp.), holly (*Ilex aquifolium* L.), and Holm oak (*Quercus ilex L. subsp. ballota* (Desf.) Samp.) [6]. Consequently, numerous tree species, generally deciduous ones, have been subjected to this traditional management regime.

In order to accurately interpret the evolution of European forests, especially in the south, anthropogenic disturbances, as well as the development thereof over time, must be taken into account. This can involve erroneous interpretations of the patterns and processes observed in forest ecosystems [7–9]. Specifically, pollarding has constituted a fundamental part of rural economies in much of Western Europe, and it also shapes a landscape of great cultural, aesthetic, and identity-related value that has lasted up to the present time as a result of the persistence of these practices [10–14]. Pollarding became widespread in many of Europe's forests and agricultural and rural environments, from the British Isles to Romania and from Scandinavia to the Mediterranean basin, and evidence of pruning in the different agro-forest landscapes can still be observed [5, 15–17].

Many studies on *Fraxinus* ssp. in Europe have been conducted in relation to the taxonomy and the ecological characteristics of *Fraxinus excelsior* L. [18–22] and to the role it plays in maintaining biodiversity [23, 24]. Nonetheless, few studies on *Fraxinus angustifolia* Vahl. have addressed the relevance of this species in the western Mediterranean's agro-forest landscape. The centre of the Iberian Peninsula houses one of Europe's densest formations of *Fraxinus angustifolia* [25]. These forests are currently conserved as *dehesas*, mainly presenting tree and herbaceous strata associated with secular livestock farming which to the present day still involves pollarding. It is precisely this *dehesa* landscape that characterises the formations of *Fraxinus angustifolia* in the Guadarrama mountains of Madrid. These dehesas formed mainly (sometimes exclusively by *Fraxinus angustifolia*) are really unique because of their main species and their territorial concentration and differ from the best-known dehesas of the South and West peninsular, usually populated by *Quercus suber* and *Quercus ilex*. Pollarding can be conducted twice during the year: in wintertime, less habitual and intended mainly for firewood production, and at the end of the hot season, when pastures are parched, in order to provide fresh fodder for the livestock immediately after pruning. This endows the tree with a unique appearance, with stems up to 2 m long which become wider at the top due to the scarring where they are cut. This pruning system gives rise to *dehesas*, dotted with old trees presenting a *cabeza de gato* (cat's head) morphology. Notwithstanding, this structure is not homogeneous because these large stands alternate with young stands and flooded areas. This agricultural landscape also comprises pastures growing in the herbaceous stratum. These formations exhibit a certain degree of variety based on gradients of moisture, with xeric species (dense, high-cover pastures dominated by *Poa bulbosa* L.), on relatively moist (pastures of *Agrostis castellana* Boiss. and Reuter) or temporarily flooded (moorlands of *Nardus stricta* L.) terrain [26, 27]. This whole mosaic, dominated by pollarded trees, makes up an open *dehesa* forest formation. It presents a high biodiversity of energy-rich pastures on flat or slightly sloped terrain and in turn constitutes one of the most emblematic traditional landscapes on the southern slopes of the Guadarrama mountains.

Interestingly, *Fraxinus angustifolia* plays a significant role in the areas close to the villages, known as *ruedos* in the Mediterranean world. This space mainly involves smallholdings delimited by stone walls and hedges, which are subjected to intensive cultivation; they are habitually irrigated and therefore permanently moist practically throughout the year. In this environment, the most common tree species is the ash (*Fraxinus angustifolia* L.). This landscape comprises hedgerows in which the trees and the ash in particular play a relevant role, both within and around the perimeter. It is a fragile landscape, endangered by intense processes, mainly urbanisation and the abandonment of agriculture and livestock farming, factors which cause the loss of much of these lands [28].

Furthermore, in these forests, many associated values are recognised, not only in relation to the biodiversity and the ecological elements they contain but also to the culture and identity-related values and those referring to aesthetics, perception, productivity, and history. Civil society promotes numerous initiatives emphasising the heritage-related values of pollarded landscapes or of unique examples of pollarding: the Woodland trust (http://www.woodlandtrust.org.uk/), the Veteran Trees Initiative (UK.), The centre Européen des trognes in France (http://www.maisonbotanique.com/centre-europeen-trognes.php), the Vetree European Project (Veteran Tree Network, https://vetree.eu/es/page/10/), etc. In Spain, we can highlight the initiative "Chopo Cabecero (Pollarded *Populus*). The Identity of a Landscape", implemented by the Centro de Estudios del Jiloca—the Jiloca Study Centre (http://www.chopocabecero.es/).

The present research aims to provide more in-depth knowledge of *dehesas* of pollarded *Fraxinus angustifolia* in the centre of the Iberian Peninsula and to elucidate the origin and genesis of these formations on the southern slopes of the Guadarrama mountains; to demarcate and map the area they occupy; to characterise and to establish the typology of the masses delimited and lastly, to identify their main values and the threats they face.

2. Methods and study area

2.1 Methods

The study was developed in four methodological phases in order to establish the area covered by *Fraxinus angustifolia*, to analyse its state of conservation and forest structure, to determine the main variables influencing its distribution and lastly, to establish a typology.

In the first phase, we conducted a search for bibliographic and digital information at different scales, and we analysed documentation in digital historical archives (Archivo PARES: http://pares.mcu.es/) and municipal archives. As sources of basic cartographical reference, we employed the Inventario Forestal de la Comunidad de Madrid—Madrid Regional Govt. Forest Inventory (Consejería de Medio Ambiente—Dept. of the Environment), the Mapa Forestal de España—Spanish Forest Map, the habitats considered in Directive 92/43/CE, and the Mapa de Terreno Forestal de la Comunidad de Madrid – Madrid Regional Govt. Map of Forestland. As complementary sources, we employed the MDE LIDAR (Instituto Geográfico Nacional—National Geographic Institute), the Madrid Regional Govt.'s available planimetry at a scale of 1:5000, the Catastro de Rústica—Rustic Land Registry (Dirección General del Catastro—General Land Registry Dept.), and the orthography at a scale of 1:5000 (IDE of the Madrid Regional Govt. available through WMS).

In the second phase, we established the criteria for identification, correction, and incorporation of enclosures containing ash trees. This enabled us to identify the ash forests that were not defined in the cartography to maintain the pre-existing ones and to eliminate those that could no longer be considered as such. We bore in mind:

- All the areas included in the Iberian Mediterranean Ash Forests of *Fraxinus angustifolia* and *Fraxinus ornus* 91B0 (Directive 92/43/CEE), eliminating riparian forests with excessively regular widths. In these areas, we verified in the field the absence of ash trees and excluded the areas dominated by riparian vegetation.

- The polygons classified as ash forests, ash tree *dehesas*, and a mixture of ash trees with deciduous species from the cartography of the Mapa de Terreno Forestal de la Comunidad de Madrid—Madrid Regional Govt. Map of Forestland. We also included polygons in which *Fraxinus* presented values of 30% or others coinciding with hedgerows and wood pastures.

- Riparian ash forests that presented continuity with larger masses or hedgerows with presence of ash trees.

- We did not take into account the riparian ash forests; however, we did establish criteria for differentiating them. For this reason, we employed the basic hydrographic network from the MDT 100 × 100 to define a 100 m buffer zone (Public Water System); we conducted an individualised review of the areas that remained within the buffer zone or those that were situated very close to it. This enabled us to include those that presented continuity with other polygons of ash forest or those that, although they formed part of the riparian vegetation, encroached uphill or had become meadows or plots of *dehesa* land.

The third phase consisted of designing different itineraries combining the orthography at a scale of 1:5000 and employing the LIDAR Digital Elevation Model (0.5 m) and the available forest cartography. We selected the tracks considering geographic area: Northeast, Centre and Southwest, and abiotic (physiography and lithology) and forest (type of forest structure) conditioning factors. These transects enabled us to rectify and incorporate new data into the cartographic information sources of reference.

In the final phase, we estimated the different variables intervening in the localisation of the ash forests, reducing these to three: slope, morphology of the terrain, and surface soil moisture. We obtained the slope directly from the LIDAR model, simplifying it in order to operate it in three categories (low: <5%; moderate: 5-10% and medium>10%). Curvature, derived from the MDE LIDAR, was classified into two categories: 1 (concave lands) and 2 (convex lands). Finally, for moisture, we differentiated three typologies: moist, semi-moist, and dry. In the latter case, we applied particular methodological criteria. We first analysed the images from Sentinel-2 in the months of May, June, and July (clear contrasts between moist and dry areas), choosing as the optimum image the one from July 3, 2015. We extracted data on surface soil moisture using the band combination that was proposed by EOS DATA ANALYTICS (2017) and reclassified into four intervals for subsequent vectorisation.

With the aim of establishing a valid typification, we analysed the three variables obtained and conducted a grouping process using the software ArcGIS 10.3 (ESRI) *Grouping Analysis* tool. We verified the reliability of the results by means of a stepwise or relational analysis with the *exploratory regression* algorithm of the same software. We calculated the results following the *euclidean distance* method, avoiding sub-algorithms, which provide excessive weight to geographic proximity. The information was simplified until four types were established. To this, we added the cohort of species accompanying each type (based upon the Madrid Regional Govt. Map of Forestland), differentiating: monospecific ash forests, with *Quercus pyrenaica* or *Quercus ilex subsp. ballota*, with different types of scrublands or mixed with different deciduous species. Finally, in order to provide an understanding of the distribution of *Fraxinus angustifolia* in the area, we intersected the types with the morphostructural units identified. In this case, as a reference we employed data from [29, 30] differentiating three groups: horst, tectonic basins and depressions and piedmont[1].

[1] Based on an initial classification in which 6 types were differentiated: massif horst and slopes, massif horst and summits, medium-sized horst and slopes, tectonic basins, depressions and piedmonts.

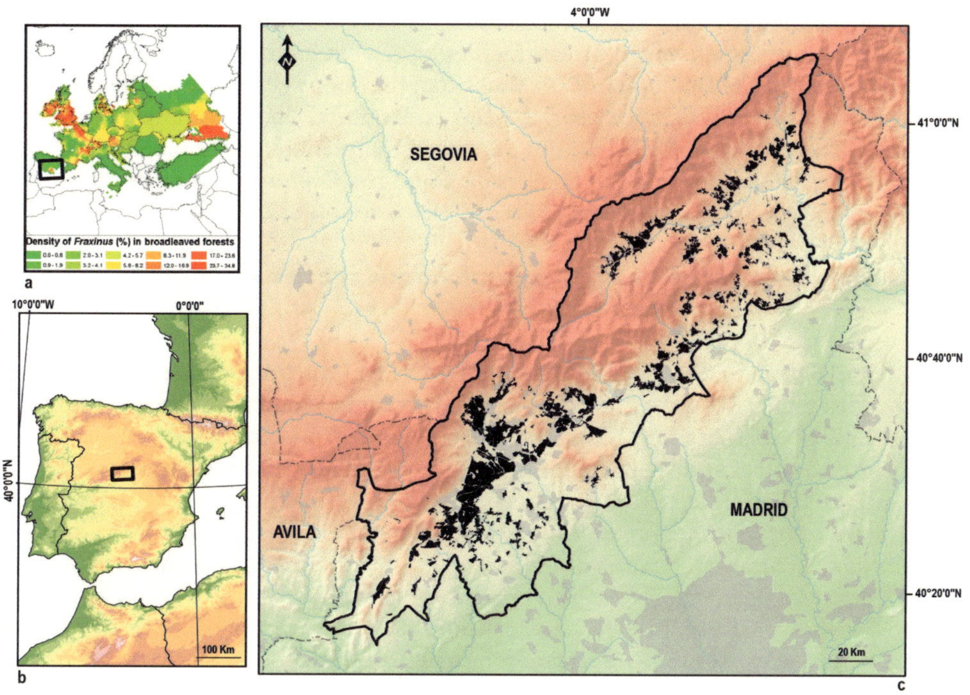

Figure 1.
a) Fraxinus angustifolia density in Europe according to [25]. b) Localisation of the study area on the Iberian Peninsula. c) Distribution of the Fraxinus angustifolia forests in the study area. Own design.

2.2 Study area

The area chosen is located on the southern slopes of the Guadarrama mountains, where *Fraxinus* exhibits one of its finest and southernmost representations in Europe and in the Iberian Peninsula (**Figure 1**). Specifically, we studied 49 municipalities presenting a total area of 198,300 ha. In this sector, the morpho-structural guidelines condition the layout of the relief, which is based on a tectonic system running in the northwest-southeast direction [29–31]. The physiography is characterised by horst (at approx. 2000 m) and graben (from 900 to 1000 m). Lying within the latter, both on the periphery and between summits are tectonic basins and depressions. The dominant materials comprise granitoids and metamorphic rock fields occasionally interrupted by calcareous-marly outcrops. The climate is continental Mediterranean, characterised by hot and dry summers, an average precipitation slightly higher than 650 mm annually, concentrated in autumn and winter, and average temperatures ranging from 12 to 14°C. In this sector, *Fraxinus angustifolia* L. occupies 20,591 ha of the more accessible zones containing tectonic basins, depressions and gentle slopes, on soils with little or no surface hydric deficit. These morpho-hydrological conditions give rise to the aforementioned rich pollarded livestock-farming silvosystems that are highly appreciated for their productivity.

3. Evolution, origin, and historical background

3.1 From the forest mass to the initial uses (eleventh to fourteenth centuries)

The gradual sedentarisation and growth of settlements during the reconquest against Islam (mid-twenty-first century) gave rise to the first production-oriented forest management based on pruning. Specifically, the location of settlements in valley bottoms occupied by ash forests facilitated the exploitation thereof, with the appearance of the first pollards. At that time, there was generally abundant regulation of uses in the shape of Royal Decrees, Regional Codes, and Byelaws. This regulatory framework enforced within a series of incipient territorial identities (the *Real de Manzanares*, the *Sexmo de Casarrubios,* and the *Comunidades de Villa y Tierra* of *Buitrago* and of *Segovia*) lasted almost five centuries [4, 6, 7, 8, 28, 32–34, 36, 37].

3.2 The coppice forest of *Fraxinus angustifolia* and the first enclosures (fourteenth to eighteenth centuries)

The population increase in the mountain areas and the progressive appearance of new settlements caused a growing need for pastures, charcoal, firewood, and timber. This entailed obvious risks and consequences: transformation of the ash forests into coppices and their delimitation as municipally property. In the High Middle Ages, a period of regulation was initiated [32], in which these delimitations were decisive: *las fresnedas son acotadas y defesadas restringiendo sus mejores vuelos y herbajes para el engorde de bueyes de labranza (the ash forests are delimited and "defesadas" (protected), restricting their best crowns and pastures in order to fatten beasts of burden)*. It was precisely this need that caused the continual modification of byelaws governing forestry uses [33–37]. Numerous examples exist of different municipal byelaws, highlighting the importance of protecting these forests from the depletive uses that were becoming generalised outside these estates. For instance, in the case of El Escorial, the Royal Decree dated September 3rd, 1565, prohibits introducing *"...ningún género de ganado mayor ni menor, ni de noche ni de día..." (any class of large or small livestock, whether by night or by day), and forbids "... sacar ni cortar ninguna leña verde ni seca del heredamiento de la fresneda..." (extracting or cutting any green or dry wood from ash forests)*. These circumstances generally applied, with few exceptions, to the whole piedmont of the Guadarrama mountains. At this time, there were two models of exploitation of the ash forests. On one hand, the *dehesas boyales* (common-use pastures), subjected to a higher degree of regulation and protection, used for livestock farming, in which one could find examples of pollarding. On the other, there were the common and private ash forests, the exploitation of which involved coppice forests exploited for charcoal and firewood.

3.3 Genesis of pollarded forests (eighteenth to nineteenth centuries)

Once the Court of Madrid has become consolidated, the increased demand for charcoal and timber determined the management model of the ash forests (eighteenth century). As from this time, the population growth gave rise to an exponential increase in the need for fuel (firewood and charcoal) and meat [38]. This demand for fossil fuels is the commonplace in the rest of Europe and has serious consequences for the forest formations [39, 40]. For example, [41] state that in Europe, wood was habitually extracted from coppice forests to make charcoal and they consider that at least 15% of this came from wood pastures. This rise in consumption of forest resources called for State intervention by means of the 1748 Royal Byelaws, which were intended to protect and increase the area of forests and plantations. During the eighteenth and ninteenth centuries and within this reference framework, the State attempted to tax these resources, conducting as many as three inventories, see [42–44]. Analysis thereof enables us to determine the presence of three different typologies associated with forestry uses in general and of ash forests in particular: coppice forests used for production of charcoal and firewood and seedling forest used for production of timber for construction and pollards in enclosed pastures intended to provide grazing for beasts of burden.

Despite the attempt of local councils to conserve forests and *dehesas*, in the nineteenth century, deforestation and overexploitation were noteworthy. This can be seen in [45]: *"Se observa bastante monte de roble, quejido y fresno aún en los terrenos labrados, lo que hace suponer que en su antigüedad estuvo cubierto de uno muy espeso y que las necesidades de cultivo y la industria del carbón, juntamente con la presión de combustible para hacer fuego los habitantes del país, han dado lugar a que sólo exista dos terceras partes o la mitad propiamente de esta clase de terreno en los alrededores de la población..." (One can observe common oak, lusitanian oak and ash forest remaining on the cultivated land, which leads one to believe that in the old times it was covered with dense forest, and the need to cultivate, together with the charcoal industry and the inhabitants' firewood requirements, has led to a situation in which only two thirds or one half remain of this kind of land in the surroundings of the village).*

Unlike other European countries, such as the United Kingdom [3], where pollarding fell into decline as from the year 1900, in our study area, it became consolidated over time and is actively practiced at present.

3.4 Generalisation of the current forestry model applied to ash forests (nineteenth to twenty-first centuries)

As from half-way through the nineteenth century, there occurred a gradual decrease in the multifunctionality of forests, with the consequent decline in their exploitation for charcoal and firewood. This change in the model was confirmed in the twentieth century, and a new production system became consolidated and has come to dominate in the pollarded ash forests: extensive livestock farming [46]. In this context, many farmers began to breed fighting bulls and from the start of the nineteenth century supplied animals for Madrid's bullfighting festivities [4]. Additionally, in view of the low profitability of bull breeding, many owners began to produce quality meat by means of imported breeds (Limousin and Charolais), autochthonous ones (Negra Serrana), or a mixture of both.

There is currently clear evidence of ash forests mainly associated with livestock farming, which presents a high degree of silvo-structural diversity. Nonetheless, the best ash forests are maintained as functional pollarded open woodland, and the marginal lands are progressively becoming pluri-specific forestland with thickets.

4. Types of pollarded ash forests and their values

4.1 General characteristics

Fraxinus angustifolia occupies 20,590.80 hectares, representing 10.4% of the study area and approximately 3% of the Madrid Regional Autonomy. These formations run parallel to the predominant reliefs in this sector of the Guadarrama mountains (**Figure 1**). Their highest degree of concentration and continuity is seen in areas presenting very specific geographic and environmental features: high moisture levels, gentle slopes and altitudes ranging from 800 to 1350 metres on granitic and metamorphic lithologies. These silvo-structures present an appearance of open woodland with pollarded trees, and they represent one of the most characteristic landscapes on the low slopes and in the valley bottoms.

Almost 80% of the ash forests mapped present a fraction of canopy cover of between 10 and 70%. Furthermore, very few are included in very dense forest

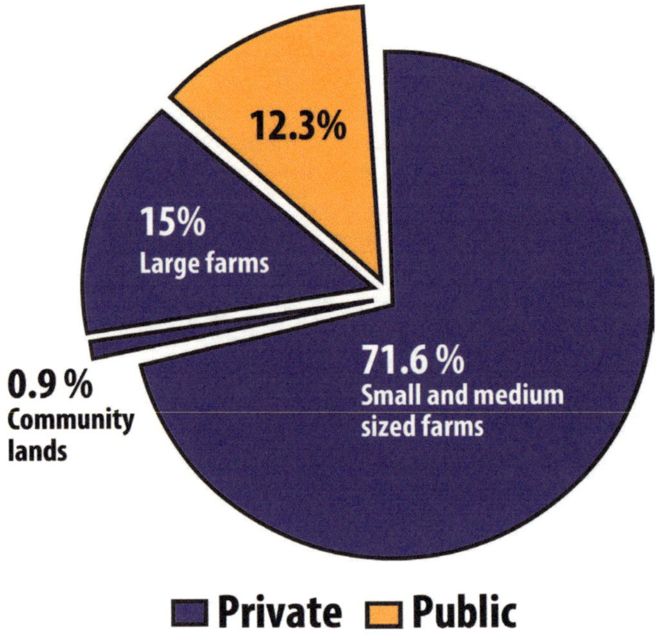

Figure 2.
Ownership of the ash forests. Own design based on the Madrid Regional Govt. Map of Forestland (2009).

formations (2.34%). The remaining 20% are integrated within meadows or pastures with dispersed trees, showing no continuity.

As for ownership (**Figure 2**)[2], there is a clear relationship between the ash forests and private property, as over 85% (18,346.85 ha) are privately owned. In particular, there is a predominance of private estates (86.70% and 18,141.21 ha), and in this category, the representativeness of neighbourhood associations falls to 0.98% (205.64 ha). Among publicly owned lands, we can highlight town councils (8.47%; 1771.80 ha), compared with 3.86% belonging to the Madrid Regional Govt., the State, and the *Confederaciones Hidrográficas* (Water Boards). Lastly, with regard to the type of property, there is a predominance of medium- and small-sized private estates (71.6%, 14,986.34).

4.2 Types of ash forests

We defined four typologies of ash forests in relation to moisture, slope, morphology of the terrain, and accompanying species (**Figure 3**, **Table 1**). As can be seen in **Table 1**, the variable moisture constitutes the most influential conditioning factor of these four types. The factors relating to slope and accompanying species are secondary and are represented in all the types and at all thresholds. In relation to the latter variable, the species and categories selected were monospecific ash forests; ash forests with *Quercus pyrenaica*; with *Quercus ilex subsp. ballota*; with different types of scrubland and with a mixture of different deciduous species. As can be observed in **Table 1**, the mixed formation of *Fraxinus angustifolia* and other deciduous species is almost insignificant in

[2] There is a discrepancy between the total area considered as ash forest and the area of variables associated with ownership (2924.69 ha) due to the inclusion of some ash forests situated outside what is considered as "Forestland" on the Madrid Regional Govt. Map of Forestland (municipal green spaces, on occasions considered as urban land, etc.).

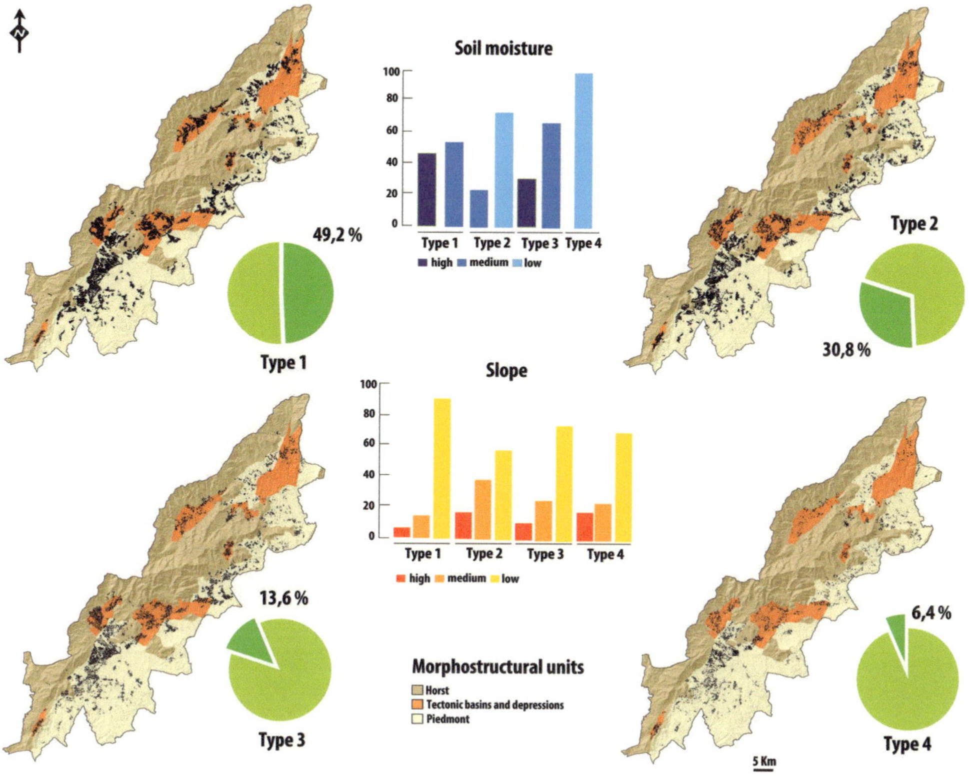

Figure 3.
Types of Fraxinus angustifolia forests.

the four types obtained, ranging from 0.56 to 0.77% (155,59 ha). The facies of *Fraxinus angustifolia* with scrubland is an indicator of the possible process of abandonment of these formations in the study area. The most habitual species involves dense thorny thickets of blackberry (*Rubus ulmifolius*), hawthorn (*Crataegus monogyna*), and wild rose (*Rosa* ssp.). The percentages are very similar in the four types obtained, ranging from 9.48 to 10.67% (a total of almost 2000 ha). Therefore, the most characteristic accompanying species involves *Quercus ilex subsp. ballota* and *Quercus pyrenaica*, which would appear to indicate transitions toward different bioclimatic belts. The presence of *Quercus ilex subsp. ballota* accompanying *Fraxinus angustifolia* is associated with the upper horizons of the Mesomediterranean and the lower horizons of the Supramediterranean, with more xeric edaphic conditions. Specifically, this formation appears in over 3195 ha and has its maximum representation in type 2 (20.03%). The main plurispecific formation therefore involves *Fraxinus angustifolia* accompanied by *Quercus pyrenaica*. In this case, it is associated with the Supramediterranean belt, with soils exhibiting a higher level of surface moisture. This formation occupies over 6400 ha, and type one concentrates over half, representing 32.48% of the area. *Fraxinus angustifolia* disappears when conditions relating to stoniness, xericity, and thermicity become more favourable to *Quercus pyrenaica* with increased altitude.

The various different types of *Fraxinus angustifolia* forests are described below. The extension and main characteristics are mentioned for each typology.

Type 1. These are mainly monospecific formations occasionally sharing the forest cover with *Quercus pyrenaica*. They are located on flat topographies, gentle and

		ha	%	ha	%	ha	%		ha	%	ha	%	ha	%
		1	1	2	2	3	3		1	1	2	2	3	3
Type 1 (10135.39 ha - 49.22%)	**Moisture**			**5645.06**	**55.7**	**4490.33**	**44.3**	Type 2 (6340.89 ha - 30.79%)	**4792.72**	**75.58**	**1548.17**	**24.42**		
	fa+qi			1225.61	12.09	129.54	1.28		998.86	15.75	270.91	4.27		
	fa+scr			514.81	5.08	446.09	4.4		462.14	7.29	140	2.21		
	fa+qp			1098.1	10.83	2193.67	21.64		1438.98	22.69	539.2	8.5		
	fa+od			10.87	0.11	67.2	0.66		41.86	0.66	7.07	0.11		
	fa			2795.67	27.58	1653.83	16.32		1850.89	29.19	590.99	9.32		
	Slope	**8779.95**	**86.63**	**984.89**	**9.72**	**370.56**	**3.66**		**3344.01**	**52.74**	**2237.6**	**35.29**	**759.28**	**11.97**
	fa+qi	1338.38	13.21	14.14	0.14	2.63	0.03		683.65	10.78	473.49	7.47	112.63	1.78
	fa+scr	866.44	8.55	69.31	0.68	25.14	0.25		290.54	4.58	207.9	3.28	103.7	1.64
	fa+qp	2486.65	24.53	586.31	5.78	218.81	2.16		819.65	12.93	795.67	12.55	362.86	5.72
	fa+od	51.02	0.5	16.42	0.16	10.63	0.1		17.52	0.28	23.06	0.36	8.34	0.13
	fa	4037.45	39.84	298.71	2.95	113.34	1.12		1532.64	24.17	737.49	11.63	171.74	2.71
Type 3 (2795.96 ha - 13.58%)	**Moisture**			**1885.76**	**67.45**	**906.56**	**32.42**	Type 4 (1318.55 ha - 6.4%)	**1318.55**	**100**	**0**		0	
	fa+qi			314.02	11.23	25.3	0.9		232.15	17.61				
	fa+scr			172.52	6.17	106.67	3.82		140.73	10.67				
	fa+qp			366.66	13.11	453.14	16.21		323.11	24.51				
	fa+od			3.45	0.12	17.78	0.64		7.36	0.56				
	fa			1029.1	36.81	307.3	10.99		615.19	46.66				
	Slope	**1874.67**	**67.05**	**650.06**	**23.25**	**271.23**	**9.7**		**867.14**	**65.76**	**266.56**	**20.22**	**184.85**	**14.02**
	fa+qi	252.07	9.02	67.47	2.41	19.79	0.71		148.73	11.28	50.38	3.82	33.04	2.51
	fa+scr	210.72	7.54	49.9	1.78	18.57	0.66		81.85	6.21	26.18	1.99	32.71	2.48
	fa+qp	424.72	15.19	260.45	9.32	134.63	4.82		136.99	10.39	96.3	7.3	89.83	6.81
	fa+od	13.05	0.47	4.25	0.15	3.93	0.14		3.2	0.24	2.66	0.2	1.51	0.11
	fa	974.11	34.84	267.99	9.58	94.31	3.37		496.38	37.65	91.05	6.91	27.77	2.11

Moisture	1: low; 2: medium; 3: high	
Slope	1: <5%; 2: 5-10%; 3: >10%	
Accompanying species	fa+qi	*Fraxinus angustifolia* + *Quercus ilex* subsp. *ballota*
	fa+scr	*Fraxinus angustifolia* + scrubland
	fa+qp	*Fraxinus angustifolia* + *Quercus pyrenaica*
	fa+od	*Fraxinus angustifolia* + other deciduous
	fa	*Fraxinus angustifolia* (monospecific ash forests)

Table 1.
Types of ash forests according to moisture, slope and accompanying species.

very gentle slopes (<5^0), fundamentally in depressed areas. A total of 49.22% falls within this group (10135.39 ha).

Type 2. This type comprises formations always presenting low or medium moisture levels, with no moist ash forests. The principal accompanying species are *Quercus pyrenaica* and, to a lesser degree when the surface is very dry, *Quercus ilex subsp. ballota*. They are located on medium and gentle slopes (<5–10^0) and on depressed terrain. They occupy 30.79% (6340.89 ha).

Type 3. Comprises the semi-moist formations and, to a lesser extent, the moist ones. There are no ash forests with low surface moisture. Almost 50% are monospecific and are situated on medium and gentle slopes. A total of 13.58% of the ash forests fall within this type (2795.96 ha).

Type 4. Comprises the dry monospecific ash forests on dry flat and sloping topographies. Only 6.4% falls within this category (1318.55 ha).

The definition of the typologies reveals the localised nature of the ash forests. As shown by **Figure 3**, the distribution of the different types in the territory studied does not clearly respond to patterns of geographic distribution because the variables influencing their localisation are present in most of the study area. For this reason, they intersected with the large morphostructural groups of the Guadarrama range. Results show that 61.09% (12,579.72 ha) of the ash forests lie on piedmont, 31.72% (6531.88 ha) on tectonic basins and depressions, and 7.18% (1479.20 ha) upon horst and its gentle slopes. We now indicate the main features of the morphostructures and the percentage distribution in hectares of each typology of ash forest in relation to the total area occupied by ash forests.

4.2.1 Piedmont

In these spaces, the predominant landscape exhibits a gentle slope, with visual basins opening up toward the Tagus or Torrelaguna river basins, with the Guadarrama mountains forming a backdrop.

At the foot of Guadarrama's main reliefs lies large areas gently sloping towards the Tagus River Basin or smaller areas sloping towards interior basins such as that of Buitrago (**Figures 4** and 5). Their morphology generally presents broad ramps that gradually disappear, sinking below marls and gypsum formations. The readjustment tectonics have determined the current configuration and layout, favouring the entrenchment of the rivers and their entombment in the more fractured areas.

Type 1 is most relevant in the piedmont: 28.87% (5943.64 ha). Its most representative elements are located in the granitic piedmont of El Escorial-Alpedrete-Guadarrama (south-western sector). In these areas, fracturing has favoured the presence of depressed sectors of different sizes where accumulation of surface and hypogeous water is relatively common. The micro-topographical conditions, particularly on the rockier piedmont, give rise to different transition facies in which the remaining minority types are included. Type 2 (20.17%, 4153.71 ha) is associated with granite rocks or more continuous outcrops. In certain discharge sectors, type 3 (7.97%, 1641.23 ha) is associated with local fractures or surface runoffs and type 4 (4.09%, 841.14 ha) colonises marginal sectors on gentle slopes or structural thresholds.

As for ownership, in this context, the large estates (generally rare), together with the small- and medium-sized ones, play a vital role.

4.2.2 Tectonic basins and depressions

This landscape presents a combination of large tectonic basins between mountains, such as the Lozoya Valley, together with small depressions, presenting rounder forms and alveolar weathering basins. In both cases, the landscapes are closed, exhibiting compartmentalised configurations, as well as a strong identity.

Tectonics plays a fundamental role in this unit (**Figures 6** and 7). The graben is characterised by constituting large sectors with irregular morphology, delimited by a well-defined tectonic scheme along all its margins. They are mostly

Figure 4.
Panoramic view of the El Escorial piedmont.

Figure 5.
Pollarded specimen in the El Escorial piedmont.

Figure 6.
An example of depression (arroyo del Valle, Bustarviejo).

Figure 7.
Group of pollarded trees (Valle del Lozoya).

situated at the foot of large horst. In the case of the depressions, the configuration adheres to fractures, presenting a certain degree of linearity and river valleys well-defined at one end (i.e., Robledo de Chavela, Manzanares el Real). It is often difficult to differentiate between types, as they can display mixed morphostructures.

Within this morphostructural unit, type 1 constitutes the most dominant one in the flatter sectors (17.52%, 3606.58 ha). Characteristic examples are the central sector of the Lozoya Valley or the depressions in the form of large alveolar weathering basins, like the arroyo of the Bustarviejo Valley. Type 2 is localised in the marginal areas in the rock fields in interior of the unit (7.97%, 1641.35 ha). Lastly, types 3 and 4 are hardly represented herein (4.46%, 918, ha and 1.78%, 365.95 ha, respectively) and are situated in marginal sectors coming into contact with the gently sloping horst.

4.2.3 Horst

They constitute the nucleus of Guadarrama's summits and slopes (**Figures 8** and **9**). Their current morphology derives from the effects of the postalpine tectonic readjustment that gave rise to a succession of horst at different levels. This unit extends from the summit erosion surfaces to where they come into contact with tectonic basins and depressions. Continuity is maintained along the axis Siete Picos-Carpetanos-Cuerda Larga, and marginal sectors are maintained in La Cabrera and in the displaced horst of Cerro de San Pedro.

Within this morphostructure, the ash forests are not very representative, appearing in the lower slope areas. Types 1 and 2 are the ones that occupy the largest area (2.84%, 585.18 ha and 2.65%, 545.82 ha, respectively). They are represented by formaciones finícolas (peripheral formations, that is, ones living at the edge of their range) that become installed in the flat areas occurring on steep slopes (i.e., Canencia) or in arroyos with stepped slopes (i.e., La Acebeda, Miraflores, Navacerrada). The presence of types 3 and 4 is not significant, as they barely reach 1.69% of the total (348 ha).

Figure 8.
Abandonment of hedgerows and dead ash tree (Canencia).

Figure 9.
Hedgerow and grazing pastures with Gladiolus communis (Lozoyuela).

	Values	Threats
Ecological and Biological	• High entomological biodiversity (saprophytes, lepidopters...) • Bird reserve • Well-conserved forest masses • Unique specimens	• Loss of entomological biodiversity associated with absence of livestock farming • Loss of traditional practices • Absence of pruning (hyper-development of crowns, risk of extreme climate phenomena, mortality...). • Intensification of, and changes in, production systems • Disappearance of stone walls • Closure of estates • Disappearance of pollarded ash trees due to the spread of urbanisation
Cultural and identity-related	• Livestock farming and pruning practices presenting a high ethnographic value • Robust identity and emotional link with these formations • Unique specimens	
Aesthetic and perceptive	• Patches of a certain extension in high-visibility environments (piedmonts and graben) • Intense annual phenological variety • Cultural image associated with artistic expression	
Productive	• Pastures with a high livestock-farming value • Estates and farms used for breeding fighting bulls. Use for firewood and timber • Apiculture • Tourism resource	
Historical	• Forms inherited from old byelaws • Conservation of stone walls and dry stone walls • Maintenance of common-use municipal estates • Presence of charcoal furnaces	

Table 2.
Values of, and threats facing, the ash forests in our study area.

4.3 Heritage values of the ash forests

Apart from characterising these ash forests, we also describe in depth the multiple values they present (**Table 2**). Among the most notable of these, habitually indicated, are the ecological and biological values, with emphasis upon the entomological biodiversity; the productive ones because, as a result of the secular livestock farming activity, the pasture formations present a high value. Moreover, cultural and identity-related values, referring to pruning practices that present a high ethnographic value, establish a close association with the resulting landscape. Finally, we describe the aesthetic-perceptive values and the historic values, both deep-rooted in the villages where these formations appear [11, 15, 17, 47, 48].

Of note is the value reached by some trees of *Fraxinus angustifolia*, not only as masses or formations but also as unique specimens due to certain features they display, such as longevity, shape, localisation, etc.

Finally, the biggest threats facing these forest landscapes refer to the disappearance thereof as a result of a radical change of uses, for example, urbanisation, with a change from rustic land to land zoned for development. Furthermore, other serious threats involve the disappearance of the key elements shaping this landscape, such as pruning, walls, enclosures, etc. We must also highlight changes in land uses, such as the intensification of livestock farming, which alters and depletes the floristic composition of pastures that are adapted to an extensive management regime involving livestock rotation.

5. Conclusions

The present study achieved its main objective: mapping, typifying and characterising the ash forest on the southern slopes of the Guadarrama mountains. From an historical perspective, they can be said to constitute formations shaped by secular uses of the canopy and of the land. As the formations were located in the areas closer to the villages, the Byelaws and Regional Codes of the *Comunidades de Villa y Tierra* (communities) or of the councils constituted an instrument for regulating uses. These regulations, which still exist in many cases, have been decisive with regard to shaping the current forest structure. Results show that the ash forests presenting the highest values due to their surface area or state of conservation are the ones located in the rocky piedmont and in the lower areas in tectonic basins and depressions. Nonetheless, it is their current localisation in the more accessible sectors that poses the main threat facing these agro-landscapes, as a result of urbanisation, with villages expected to grow in the future.

Another characteristic of these ash forests involves the traditional pollarding method, a land use described in the historical documentation from the end of the eleventh century. This practice constitutes one of the most significant identity-related values of traditional highland culture in Madrid's Guadarrama mountains. As a result of the notable changes that have taken place in these sectors, a process we have already described in previous studies [49, 50], they should be considered as endangered landscapes. A specific figure of protection should therefore be proposed for these forest formations (i.e., cultural landscape); this status should be reinforced through inclusion in the regional and local planning.

Additionally, results confirm that *Fraxinus angustifolia* depends strongly upon moisture and sub-surface water accumulation and, to a lesser degree, on slope. Within a scenario of higher temperatures, with less hydric availability and greater evapotranspiration [51], the presence of water proves to be determinant in relation to the vulnerability of this species in a future context of climate change. This fragility ought to constitute a priority line of future research, addressing the species within a broader framework as an indicator of change at regional level. In turn, there is a need for models and implementation strategies adapted to the new climatic scenarios; these should involve the collaboration and support of the different stakeholders.

In conclusion, there is a need to question the current state of affairs and to propose management criteria for Guadarrama's ash forests. The latest initiatives developed relating to pollarded ash forests (Seminars on pollarded ash forests organised by Madrid's Politécnica University in November 2017 and by the Autonoma University of Madrid in November 2018) have focussed on detecting the dynamics and threats involved and on putting forward proposals to promote future maintenance of these formations. These potential guidelines are based upon three axes: establishment of regulations on pruning methods by the regional administration involving the participation of all the stakeholders (especially in terms of the interaction between forest rangers and landowners); promotion of research and knowledge in order to provide better management and actions for dissemination, awareness, and revitalisation of pollarding and of pollarded forests.

Acknowledgements

This research was funded by the Madrid Regional Govt. via a call for environmental activities and projects (Order 3326/2017), through the Project "Mejora de la cartografía del Hábitat de Interés Comunitario 91B0 Bosques de Fresnedas de *Fraxinus angustifolia* y *Fraxinus ornus* en el territorio de la Comunidad de Madrid". ("Enhancement of the cartography pertaining to the Habitats of Community Interest 19B0 Ash Forests of *Fraxinus angustifolia* and *Fraxinus ornus*") within the territory of the Madrid Regional Autonomy".

Author details

Fernando Allende Álvarez, Gillian Gomez-Mediavilla* and Nieves López-Estébanez
Department of Geography, Universidad Autónoma de Madrid, Cantoblanco, Spain

*Address all correspondence to: gillian.gomez@uam.es

References

[1] Rackam G. Woodlands. London: Collins; 2006

[2] Rotherham ID, Ardron PA. The archaeology of woodland landscapes: Issues for managers based on the case-study of Sheffield, England and four thousand years of human impact. Arboricultural Journal. 2006;**29**(4):229-243

[3] Petit S, Watkins C. Pollarding trees: Changing attitudes to a traditional land management practice in Britain 1600-1900. Rural History. 2003;**14**(2):157-176

[4] Gomez G, López N, Allende F. Las fresnedas trasmochadas del piedemonte del Sistema Central en Madrid (España): cambios y usos actuales. Estudios Rurales. 2016;**6**(11):32-47

[5] Agnoletti M. Italian historical rural landscapes: Dynamics, data analysis and research findings. In: Agnoletti M, editor. Italian Historical Rural Landscapes. Netherlands: Springer; 2013. pp. 3-87. DOI: 10.1007/978-94-007-5354-9

[6] Allende Álvarez F, Gómez Mediavilla G, López EN, Sobrino García J. Podas y trasmochos en las Ordenanzas forestales del Sistema Central español y su impronta en el paisaje forestal actua. Sociedad Española de Ciencias Forestales. 2013;**38**:35-42

[7] López Estébanez N, Gomez Mediavilla G, Madrazo García de Lomana G, Allende Álvarez F, Sáez Pombo E. The evolution of Forest landscapes in Spain's Central Mountain range: Different forests for different traditional uses. In: Rotherham I, editor. Cultural Severance and the Environment. Volume 2. Dordrecht: Springer; 2013. DOI: 10.1007/978-94-007-6159-9_11

[8] Allende Álvarez F, Sánchez MF, Mediavilla GG, Pellejero RG, Estébanez NL, De Lomana GMG, et al. Una aproximación al análisis comparativo de los paisajes forestales de la Cordillera Cantábrica y el Sistema Central. ERIA. 2014;**94**:161-182

[9] Bürgi M, Gimmi U, Stuber M. Assessing traditional knowledge on forest uses to understand forest ecosystem dynamics. Forest Ecology and Management. 2013;**289**:115-122

[10] Oria JA. Conceptos fundamentales y terminología técnica asociada a la masa forestal. In: Parketxeak G, editor. Árboles trasmochos: Tradición, gestión y conservación. Gipozcoa: Asmoz; 2011. 40 p

[11] Mansion D. Les trognes, l'arbre paysan aux mille usages. Rennes: Editions Ouest- France; 2010. 144 p

[12] Haeggström CA. Les trognes dans l'art. In: Dumont E, editor. Les trognes en Europe. Rencontres autour des arbres têtards et des arbres d'emonde. Boursay: Maison Botanique; 2007. pp. 134-140

[13] Grove AT, Rackham O. The Nature of Mediterranean Europe: An Ecological History. Yale: Yale University Press; 2003 384 p

[14] Read H. Veteran Trees: A Guide to Good Management. Peterborough: English Nature; 2000. 176 p

[15] De Jaime Lorén C. Distribución geográfica, estimación de la población y caracterización de las masas de chopo cabecero en las cuencas del Aguasvivas, Alframbra, Huerva y Pancrudo [thesis]. Zaragoza: Universidad de Zaragoza; 2015

[16] Read H. Bref tour d'horizon des arbres têtards et des pratiques d'étêtage en Europe. In: Dumont E, editor. Les trognes en Europe. Rencontres autour des arbres têtards et des arbres

d'emonde. Boursay: Maison Botanique; 2007. pp. 12-18

[17] Read H. Pollards and pollarding in Europe. British Wildlife. 2008;**19**:250-259

[18] Alhberg A. The Influence of Thinning Intensity on Stands of European Ash (*Fraxinus excelsior* L.) Affected by Ash Dieback – How Should they be Managed? [Thesis]. Alnarp: Swedish University of Agricultural Sciences; 2014

[19] Gérard PR, Fernández-Manjarrés JF, Bertolino P, Dufour J, Raquin C, Frascaria-Lacoste N. New insights in the recognition of the European ash species *Fraxinus excelsior* L. and *Fraxinus angustifolia* Vahl as useful tools for forest management. Annals of Forest Science. 2006;**63**(7):733-738

[20] FRAXIGEN, editor. Editor. Ash Species in Europe: Biological Characteristics and Practical Guidelines for Sustainable Use. UK: Oxford Forestry Institute, University of Oxford; 2005 128 p

[21] Hofmeister J, Mihaljevič M, Hošek J. The spread of ash (*Fraxinus excelsior)* in some European oak forests: An effect of nitrogen deposition or successional change? Forest Ecology and Management. 2004;**203**(1):35-47

[22] Marigo G, Peltier JP, Girel J, Pautou G. Success in the demographic expansion of *Fraxinus excelsior* L. Trees-Structure and Function. 2000;**15**(1):1-13

[23] Lõhmus A, Runnel K. Ash dieback can rapidly eradicate isolated epiphyte populations in production forests: A case study. Biological Conservation. 2014;**169**:185-188

[24] Mitchell RJ, Beaton JK, Bellamy PE, Broome A, Chetcuti J, Eaton S, et al. Ash dieback in the UK: A review of the ecological and conservation implications and potential management options. Biological Conservation. 2014;**175**:95-109

[25] Skjøth CA, Geels C, Hvidberg M, Hertel O, Brandt J, Frohn LM, et al. An inventory of tree species in Europe—An essential data input for air pollution modelling. Ecological Modelling. 2008;**217**(3-4):292-304. DOI: https://doi.org/10.1016/j.ecolmodel.2008.06.023

[26] San Miguel A, Barbeito I, Perea R, Roig S, Rojo MR. Tipología y valoración de los pastos naturales herbáceos de la Comunidad de Madrid. Pastos. 2012;**42**(1):5-25

[27] Grijalbo Cervantes J, editor. Vegetación y flora de Madrid. Madrid: Náyade Editorial; 2010

[28] Gómez Mediavilla G, López Estébanez N, Allende Álvarez F. Evolución del paisaje cultural de los campos cercados en las depresiones del piedemonte del Sistema Central (Madrid, España): modificaciones en las relaciones del espacio urbano -rural. In: Actas del XXV Congreso de la Asociación de Geógrafos Españoles. Naturaleza, Territorio y Ciudad en un mundo global. Madrid: Universidd Autónoma de Madrid; 2017. pp. 172-180. DOI: 10.15366/ntc.2017

[29] Sanz Herráiz C. El relieve del Guadarrama oriental. Madrid: Consejería de Política Territorial; 1988

[30] Bullón Mata T. El Guadarrama Occidental trama geomorfológica de un paisaje montañoso. Madrid: Consejería de Política Territorial; 1988

[31] Pedraza Gilsanz J. Sistema Central. In: Elorz G, del Olmo D, editors. Geomorfología de España. Alcorcón, Madrid: Rueda; 1994

[32] Ladero Quesada MA, Galán Parra I. Las Ordenanzas locales en la Corona de Castilla como fuente histórica y tema de investigación (siglos XIII al XVIII).

Revista de estudios de la vida local; **217**:85-108. ISSN: 0034-8163

[33] Jusdado Martín J. Las Ordenanzas, versión íntegra y literal, otorgadas en 1575 por Felipe II a la villa de Colmenar Viejo como fuente documental para su historia. Cuadernos de Estudios. 1996;**8**:9-39

[34] Aragón Nogales G. El Real de Manzanares. Diálogos de la Historia de una Comarca. Cuadernos de Estudios. 2001;**15**

[35] Ediciones Equipo A, editor. Chozas de la sierra: La construcción del espacio del agua en Soto del Real (Madrid). Madrid: Ediciones Equipo A; 2012. 276 p

[36] Sánchez Meco G. El Escorial: De Comunidad de Aldea a Villa de Realengo. El Escorial: Ayuntamiento, Área de Cultura; 1995. 709 p

[37] Fernández García G. Buitrago y su Tierra (Algunas Notas Históricas). Madrid: Imprenta Héroes; 1996

[38] López Martínez AL. Ganaderías de lidia y ganaderos. Historia y economía de los toros de lidia en España. Sevilla: Fundación Real Maestranza de Caballería de Sevilla y Universidad de Sevilla; 2002. 459 p

[39] Jansen D, Nelle O. The Neolithic woodland–archaeoanthracology of six funnel beaker sites in the lowlands of Germany. Journal of Archaeological Science. 2014;**51**:154-163

[40] Müllerová J, Szabó P, Hédl R. The rise and fall of traditional forest management in southern Moravia: A history of the past 700 years. Forest Ecology and Management. 2014;**331**:104-115

[41] Bernardos Sanz JU et al. Energy consumption in Madrid, 1561 to c. 1860. In: Massard-Guilbaud G, Mosley S, editors. Common Ground: Integrating the Social and Environmental in History. Newcastle upon Tyne: Cambridge Scholars; 2011. pp. 316-339

[42] Catastro de Ensenada. PARES Portal de Archivos Españoles. Ministerio de Cultura y Deporte. Available from: http://pares.mcu.es/Catastro/ [Accessed: 2018/10/15]

[43] Diccionario geográfico estadístico de España y Portugal, dedicado al Rey nuestro Señor/por el Doctor Don Sebastián Miñano, individuo de la Real Academia de la Historia y de la Sociedad de Geografia de Paris. Biblioteca Digital Hispana. Biblioteca Nacional. Available from: http://bdh-rd.bne.es/viewer.vm?id=0000001624 [Accessed: 2018/10/15]

[44] Diccionario geográfico estadístico histórico de España y sus posesiones de Ultramar, por Pascual Madoz. Biblioteca Digital Hispana. Biblioteca Nacional. Available from: http://bdh-rd.bne.es/viewer.vm?id=0000176537&page=1) [Accessed: 2018/10/15]

[45] Report on the forestland of Chozas de la Sierra. Toledo: Archivo Histórico de la Nobleza, Sección de Osuna; 1872

[46] Sáez Pombo E. Montes públicos, territorio y evolución del paisaje en la Sierra Norte de Madrid. Madrid: Consejería de Medio Ambiente; 2000

[47] Rush MJ. Veteran Trees Iniciative. Historical and Cultural Aspects. Northminster House, Peterborough: English Nature; 1999. 318p

[48] Woodland Trust, editors. Ancient Trees Guide n 4: What Are Ancient, Veteran and Other Trees of Special Interest? Lincolnshire: Woodland Trust; 2002

[49] De las Heras Puñal P, Fernández Sañudo P, López Estébanez N, Roldán Martín MJ. Territorial dynamics and boundary effects in a protected area of

the Central Iberian Peninsula. Open Geosciences. 2011;**3**(1):1-11

[50] López Estébanez N, Allende Álvarez F, Fernández Sañudo P, Roldán Martín M, De las Heras Puñal P. Cartography of landscape dynamics in Central Spain. In: Bateira C, editor. Cartography: A Tool for Spatial Analysis. Rijeka, Croatia: Intech; 2012

[51] CORDEX. Coordinated Regional Climate Downscaling Experiment. Plataforma sobre adaptación al cambio climático en España. Ministerio para la Transición ecológica. (2019). Available from: https://www.adaptecca.es/ [Accessed: 2018/12/22]

Chapter 4

The Portuguese *Montado*: A Complex System under Tension between Different Land Use Management Paradigms

José Muñoz-Rojas, Teresa Pinto-Correia, Martin Hvarregaard Thorsoe and Egon Noe

Abstract

The *Montado* is a silvopastoral system and the dominant land use in Alentejo (Portugal). Its functionalities expand beyond agricultural and forestry production. It is a system where a low-density and heterogeneous tree cover is complemented with livestock grazing and fodder production, resulting in its recognition as a high-nature-value farming system. However, for it to be effectively preserved, a balance between its many components needs to be secured. Despite the relevance and urgency of its conservation, the *Montado* has long suffered a constant decay. To better understand such decay, it is crucial to unravel why and how land use management decisions are made, and the interplay of drivers influencing such decisions. We applied discourse analysis to identify the various management paradigms that currently co-exist underpinning strategies by land managers and others. Our analysis is based on a review of the scientific literature, a media analysis, participant observations, and in-depth interviews with *Montado* farmers in Central Alentejo between 2014 and 2017, along with a survey with producers implemented during 2018. We conclude that existing strategies, and underpinning paradigms, are frequently incompatible, leading to the poor progress in halting the current decay of the system, and thus, also in securing its sustainability.

Keywords: *Montado*, paradigms, management, discourses, land use

1. Introduction

Agricultural production in Europe has been shifting between different management paradigms over the past 50 years more rapidly than ever before in history. Currently, multifunctionality [1] and sustainability [2] are leading the public and political agendas in the transition from productivism toward postproductivism. Nevertheless, such contemporary management paradigms have neither evolved linearly nor followed mutually successive patterns, but in much more complex and intertwined ways. Such complexity is being influenced by mutually opposing trends, such as intensification vs. extensification or global vs. local, and expands across multiple spatial and temporal scales, resulting in the current complex mosaic of rural land use and farming systems across Europe [3].

Diverse discourses concerning different farming management paradigms underpin such mosaic [4] that also overlay each other in similarly complex temporal and spatial patterns. Within this complex overall picture, some regional contexts, and thus also their farming and land use systems, are particularly vulnerable to degradation and are being affected by management and policy decisions that support the prevalence of unsustainable management paradigms. These are paradigms that too frequently disregard the complexity entailed by postproductivism at the expense of the efficiency and short-term financial returns that may be achieved through productivism [5]. This is clearly the case of Mediterranean region, where traditional rural land uses with access to key production factors (soils and water) are undergoing processes of rapid transformation through technological, ecological, and financial intensification, while those in marginal areas remain alternatively unchanged, frequently leading to their marginalization or even abandonment.

The Alentejo (Portugal) is one predominantly rural region where such a trend is clearly taking place, including in the *Montado*. This is problematic because the persistence of mutually contradictory management paradigms hampers efficiency, potentially driving environmental degradation and unsustainable land use. Despite being reflective of the above-described problems, and of its recent gradual qualitative and quantitative decline [6], the *Montado* still dominates the land use and landscapes of Alentejo (**Figure 1**).

In this chapter, we examine the range of management paradigms that currently co-exist in the *Montado*, the discourses that underpin each of these paradigms, and the tensions that arise from such co-existence. To achieve this, we first need to clarify our interpretation of what a management paradigm is. In this paper, we define a management paradigm as the technical, institutional, and legal setups influencing decisions by land use actors at the farm level, including their values, goals, beliefs, and worldviews. Thus, paradigms shape the ways in which farmers, land owners, and managers consider the *Montado* and also how they consider themselves positioned in relation to the system. Paradigms, therefore, strongly influence decisions and may potentially help explain trends in land use and management. Our aim is to unravel the diversity of management paradigms that jointly impact the future sustainability of a silvopastoral system, the *Montado*, by influencing everyday management decisions undertaken at the farm level. Following Foucault [7], we have also adopted and applied the concept of discourse to explore the underpinning reasons behind such management paradigms.

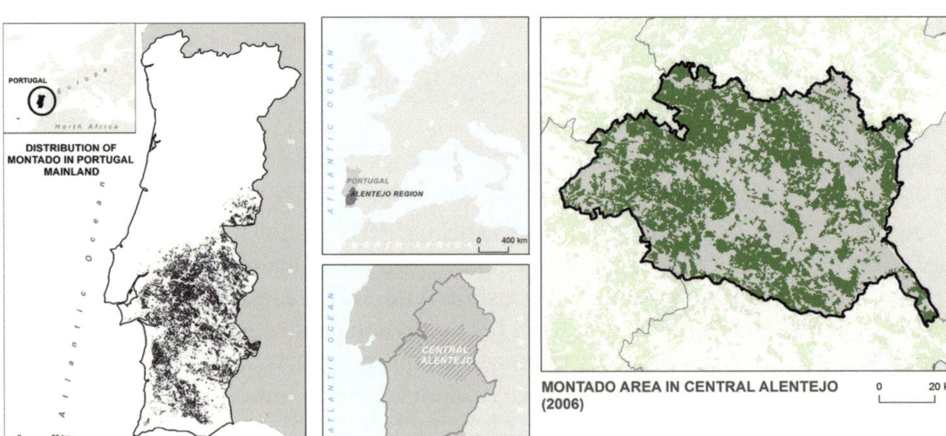

Figure 1.
Distribution of Montado in Continental Portugal (left, in black) and Central Alentejo (right, in green).

The chapter begins by describing the key structural, spatial, and socioecological characteristics of the *Montado* as it stands today, its components and their importance, the recent trends, evolution and the changes, and the main threats and opportunities. Within this initial section, we focus on two main aspects of the *Montado* that require further reflection: its current qualitative and quantitative decay, including its root causes, drivers and consequences, and its multifunctional nature, and the challenges ahead posed by intensification, globalization and financialization. Once the system has been characterized, in the third section of the chapter, we identify, describe and critically assess the multiple changes and challenges that are driving the wide range of management options undertaken in the *Montado* nowadays.

Emphasis in the chapter is made on the tensions among drivers of change across scales, including socioeconomic, financial, political and environmental ones, and hints at how this may be affecting the system's sustainability. The fourth section of the chapter identifies and explores the diverse management paradigms that can be unraveled by looking at the discourses that over the last century have dominated in the *Montado*, and how they represent different understandings of issues relevant for sustainability including power imbalances, collective perceptions, and institutional underpinnings. In the discussion, we then hint as potential pathways to progress beyond the current (problematic) situation toward more sustainable management pathways, and then close the chapter by recapitulating on the main relevant lessons learnt and barriers to progress beyond the state of the art.

2. The *Montado*

Montado is the Portuguese term used to designate the silvopastoral system, which in Spain is named as *Dehesa* [8, 9]. Montados and Dehesas spread across the central and southern regions of the Iberian Peninsula, where the dominant climate is warm-summer Mediterranean, have poor and shallow soils and strong rainfall and temperature irregularities [8, 9]. The *Montado* occupies ~1.2 million ha [6, 10], much of it in the central municipalities of Alentejo (**Figure 1**). The property structure is largely dominated by large-family-owned estates, normally ranging between 200 and 2000 ha. Only around smaller towns, a small-scale property structure appears, with a mosaic of multiple Mediterranean cultures substituting the *Montado*.

The *Montado* is characterized by a low-density tree canopy dominated by Holm and Cork Oaks (*Quercus suber* and *Quercus rotundifolia*, respectively), which can be intermixed in some areas with other tree species, including Ash (*Fraxinus* spp.) occupying the most humid soils, chestnut trees (*Castanea sativa*) in wetter and mountain climates, and other largely riparian species in the proximity of river courses, with natural or cultivated grassland in the undercover [10]. The main use of such grassland is for extensive livestock grazing, with herds of sheep or goats and Iberian pigs or cattle, depending largely on the type of terrain, soils, and characteristics of the pastures. While a few farm units breed mixed livestock, a majority of them focus only on one species, increasingly in cattle. The extensive nature of the grazing activity is justified by the strongly limiting biophysical factors, including shallow and poor soils and semidry climate, which do not generally allow for intensive grazing or any alternative or intensive agricultural land use [10].

Nevertheless, and despite a vast majority of the grazing being extensive, supplement feed is required in almost all *Montado* farms, especially during summer and autumn. Depending on management practices, pasture can be more regularly distributed or alternatively intermixed with dispersed patches of shrub. Shrub control is generally applied using livestock, often in combination with mechanical methods that are useful at certain periods of the year. Cultivation or improvement of pastures is also a common practice [8, 9].

The balance among all these components requires detailed and comprehensive management schemes, which have been improved since the eighteenth century, with knowledge being transmitted along generations, mainly within the families of *Montado* land owners and also land managers [11, 12]. Along the twentieth century, cultivation of cereals in a large rotation with pastures and fallow was common in the *Montado*, but has progressively reduced its size and is upmost rare in the present day *Montado* [13]. This is, however, in contrast with some areas of Spanish *Dehesa* (**Figure 2**), where rotations between grasslands and extensive cultivations still persist.

This is particularly the case of some plain areas in the high plateaus of Extremadura, Andalucía, and Castilla, with low-density tree canopies that allow for mechanized cropping. Although the aim of this chapter is not to deliver a comparative analysis of diverse regional situations, these cross-regional remarks might come useful to the reader, especially to those readers that are acquainted with diverse regional realities, but that need to become aware of how influential the management practices and the cultural and policy drivers behind them are. *Montados*, considered as human-shaped ecosystems, are characterized by strong habitat heterogeneity

Figure 2.
Contrast between a Dehesa in Extremadura (left) with cereal cropping in the undercover and a Montado with grazing pastures in the Alentejo (right).

Figure 3.
Landscape mosaic including land use patches of Montado with different tree densities and spatial structures, alternating with crops (mainly in the more fertile and plain areas), shrubland (in the hilliest and least productive areas), natural and artificial water masses, and other human land-cover and land use types. Overall, this should be considered a human landscape, albeit one with high levels of ecological qualities.

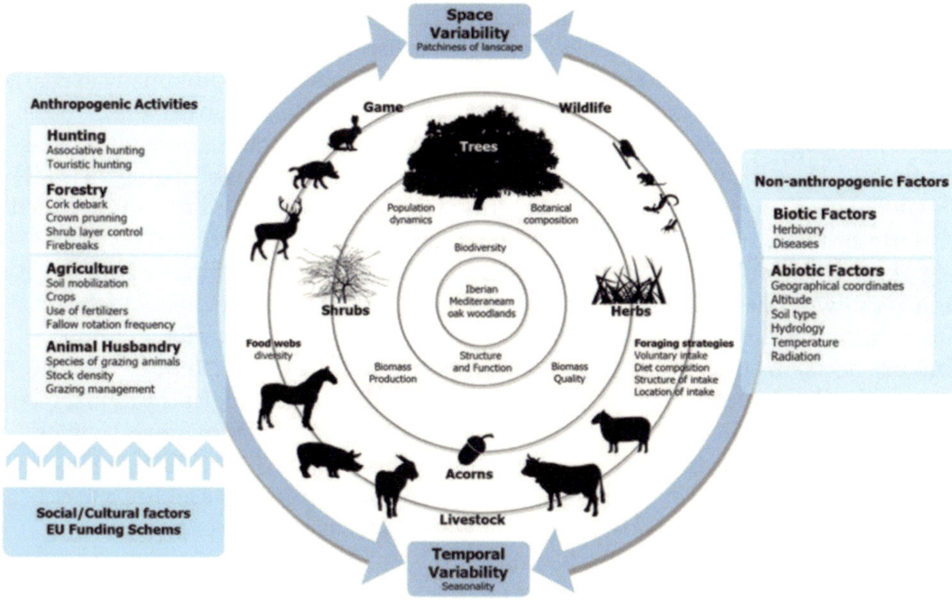

Figure 4.
The Montado as a complex dynamic land use system with its different interacting components (from [11]).

rates (**Figure 3**). This is due to the changing composition and density of the tree cover, in combination with an undercover mosaic of grasslands with dispersed or patches of shrubs. A pronounced patchiness of the vegetation communities and a marked seasonality of plant and animal biological cycles are characteristic features of *Montados* [11] (**Figure 4**). Consequently, the *Montado* has long been acknowledged as a land use system with high natural and social values, providing relevant ecosystem and landscape services well beyond the biodiversity conservation.

The specific and extremely heterogeneous environmental conditions of the Mediterranean region have played a key role in securing that these systems could prosper historically. Nonetheless, the system is now under severe threats, and undergoing qualitative and quantitative decay [6]. This is either due to insufficient innovative and adaptive land-management practices or due to pressure for rapid intensification and change. While the former trends are leading to extensification and abandonment in the less fertile and more peripheral areas, the latter are driving chronic overuse in other more fertile ones [13].

3. Ongoing changes and management challenges today

3.1 Some preliminary considerations

To critically address the key management changes and challenges being faced by the diverse key actors in the *Montado* in the early twenty-first century (farmers, managers, and policy makers), a first step is to identify and characterize the various drivers and components of the degradation trends currently affecting the system's structure and qualities. Till date, certain scientific studies have jointly assessed the spatial, ecological, and socioeconomic aspects of the degradation of the *Montado*, although this has been mostly done from a quantitative perspective, as latest assessments reflecting the state of the art seem to indicate [6]. We consider this to be clearly insufficient to provide with a full picture of the situation. Furthermore, the current trajectories of change undergoing in the *Montado* cannot be fully understood without considering the root causes and complex drivers of land use change that are set across scalar levels that largely exceed the farm units and its immediate context. These root causes generally fall completely out of control of farmers, land managers, and other key decision makers traditionally in charge of managing changes and strategies for the *Montado*, and include globalization, financialization of agriculture and liberalization of international agricultural trade, climate change and desertification (both biophysical and human) as well as changing diets [11].

Global challenges are especially problematic for traditional and multifunctional land use systems, such as the *Montado*, where the multiple services and benefits (e.g., ecological, cultural, and socioeconomic) that they have traditionally delivered are now rendered as inefficient in a political economic context on which financial competitiveness is increasingly turning into the key criterion guiding land use decision-making [8, 9]. In relation to spatial scales, we also find the scale mismatches hampering sustainability across multiple socioecological systems to be relevant for the *Montado*. This is a system on which a clear scalar mismatch exists between the land management structures and levels in place (basically focusing on the farm level) and the financial (global), institutional (National and European) and ecological (landscape) levels at which governance is actually exercised.

3.2 Ecological, spatial, and structural changes and challenges

According to recent data [6] published following a critical overview of the problems and challenges encountered in many of the datasets of official and public information, the increasing trend in the qualitative and quantitative degradation of the *Montado* is self-evident. This includes a decrease in the total size and tree density and diversity, and also in the system's overall health. According to the sixth National Forest Inventory of Portugal [14], *Quercus suber* and *Quercus rotundifolia Montados* occupy in Portugal a total of 1,067,954 ha, of which 736,755 ha correspond to *Quercus rotundifolia* and circa 331,179 ha to *Quercus suber*. Also according to [6], between 1990 and 2006, approximately 90,054 ha disappeared in the *Montado* area in the Alentejo region, with an estimated annual regression rate of 0.14% year^{-1}. This challenges the official data from the National Forest Inventory [14], which point out a relative stability in the *Montado*'s surface over the same period. Furthermore, when looked at the landscape scale, spatial data can be retrieved from official sources [14] indicating to the fragmentation and consecutively reduced ecological and functional connectivity of the *Montado*. This is clear from the gradual timely increase in the number of patches and decrease of their surface in the Alentejo region and the Portuguese region with the biggest concentration and best preserved *Montados* (*Montado* occupies 40% of the region's land cover). Regarding the plant health of the system, ICNF verified that more than 50% of the *Montado* with *Quercus rotundifolia* and 68% with *Quercus suber* were characterized by symptoms of slight decline, while in 4 and 10%, respectively, of the aforementioned systems, these symptoms are considered severe [10, 13]. Nonetheless, the *Montado* area decrease and health status should be linked with the stand development stage, as otherwise the results can be misleading.

3.3 Socioeconomic and governance changes and challenges

Generally speaking, socio-economic aspects of the *Montado* are following the same declining trends as the spatial, structural and ecological ones (**Table 1**). A 2018 survey with over 150 *Montado* farmers and farm managers has shown that only farm businesses with a minimum of 300–400 ha are now financially sustainable, a size that has grown steadily over the past few years [15]. According to these same producers surveyed, structural changes in the diverse components of the system have been happening over the past decades that have contributed to shifting its functionality, and

	1910	1960	1990	2006
Total area (km^2)	3152.95	4030.35	3544.15	3466.77
Relative area (% Central Alentejo)	43.60	55.81	49.16	47.68
Number of patches	116	208	248	306
Mean patch size (km^2)	27.18	19.38	14.29	11.33
Variance (%)	5.54	4.41	3.71	3.35
Maximum patch size (km^2)	1838.86	2496.06	2019.46	1987.46
Minimum patch size (hectares)	0.33	0.93	0.41	0.27

Table 1.
Evolution of the distribution of Montado in Central Alentejo (1910–2006), according to baseline data in [6].

ultimately, its sustainability. These include a decline in tree density (now frequently below 120 trees/hectare) accompanied by a move toward increasingly specialized cattle-focused *Montado* farms (a 2.77% increase between 1999 and 2009). This has resulted on a less diverse and resilient livestock herd, which has worsened even further by the gradually increasing trend toward the concentration of farm property [15]. Although a tree density of 100–120 trees/ha for a *Montado* can be considered as adequate from an ecologically functional standpoint, especially if crown cover is around 50–60% and if it has more than one cohort, a significant number of the 150 *Montado* producers that were surveyed [15] considered that the decline noted in tree density in many *Montado* areas should be considered not only as a proxy of the system decline, which is partly related to its decaying multifunctional character and trend toward monofunctional specialization, but also as a possible outcome of the increasingly impacting tree pests and diseases affecting the system.

Alas, other negative trends have been detected in the system that include the ever-lowering levels in market power by farmers and managers [16], as signaled by the poor rates of farmers' willingness to join, and associations noted the declining number of employments provided per farm unit. This can be linked to both the gradual increase in the relative cost of living and wages, and the intensification of the regional agricultural alternatives, which provide better paid and more specialized jobs. A last important trend detected is the human desertification occurring in the major regions on which *Montado* persists. Population density in the regions covered by Montado range between 168.1 inhabitants/km^2 in the disctrict of Setúbal [17–20] which is closest to the Metropolitan region of Lisbon to the 14.9 Inhabitants/Km2 of the Beja district [17–20]. Whilst the former is closest to the Metropolitan region of Lisbon, and thus occupies few of its human resources in agriculture, the latter is amongst the most remote and least industrialized regions in Portugal.

This results in an overall picture that indicates a difficult situation and seemingly poor prospects for the Portuguese *Montado* in the early twenty-first century. However, this problematic situation has not yet resulted in a more effective and better-coordinated responses and strategies, neither political nor social or economic, to tackle many problems of degradation signaled. It has been argued that there are cultural and political reasons behind such an inefficient response [11], but the faults and challenges in the system have not been sufficiently analyzed.

To exemplify the problems in the regulation and planning of the *Montado*, one only needs to look at the sets of legislative instruments specifically targeting the *Montado* in the Alentejo (**Figure 5**). Despite this being a region where the *Montado* is publicly acknowledged to play a key role for sustainable territorial development [18–20], the existing regulations are either partial (only protecting individual components of the system such as the oak trees [21]), insufficiently explicit and differentiated [18–20], and even advocate the support of pathways toward intensification

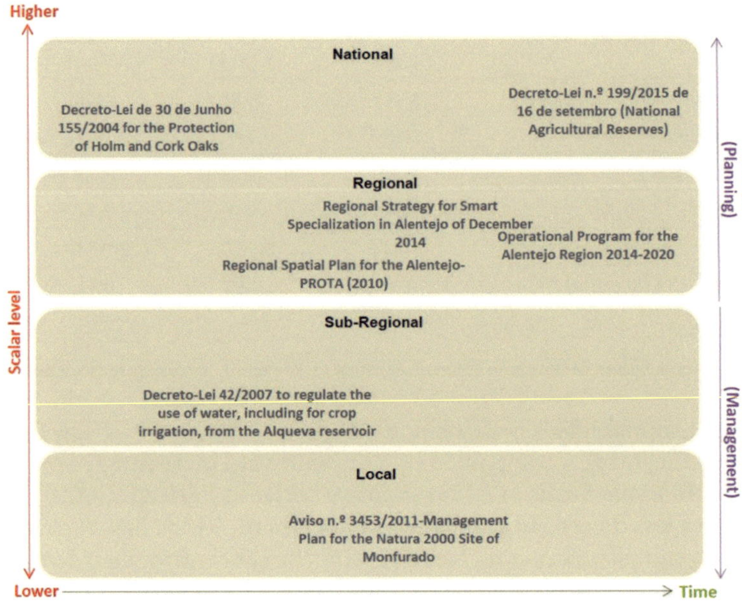

Figure 5.
Overview of the relative spatial-temporal scales and nature of the key regulatory instruments for the management of the Montado in the region of Alentejo (Portugal).

and specialization of agriculture [22]. Following such a pathway would most probably hamper the sustainability and competitiveness of traditional, extensive, and multifunctional systems, such as the *Montado*, that do not fit tightly with the "exclusivity-zoning-oriented" [23] and economic-growth-oriented approaches that have so far been preeminent in rural policies and plans across Europe.

Alas, conflict may also arise between National regulations, such as the Decreto Lei 199/2015 (National Agricultural Reserves) [24] that promote the increase of agricultural production, and other legal and planning instruments that focus on conservation of cultural, ecological, or landscape values incompatible with the intensification and expansion of crops. This is indeed the case, for instance, of the Natura 2000 network, which explicitly recognizes the urgency to promote conservation of *Montado* landscapes and habitats that are either significant or especially valuable, such as the Monfurado site in Alentejo [25]. Although in this case, zoning approaches may help prevent incompatible management paradigms (e.g., agricultural reserves and biodiversity hot-spots) to collide with time and space, they may also be considered as inefficient in light of what has been termed as the "island protection" model, which has now been long criticized as inefficient to drive change toward sustainable development [26]. Under this "island" approach, individual and frequently isolated spots are overprotected, while the rest of the territory is prone to development of different intensities, leading to a somehow "schizophrenic" territorial policy framework.

4. Examining the management paradigms in the *Montado* through discourse analysis

4.1 Management paradigms and underpinning social discourses

In order to identify and critically examine the management paradigms that have prevailed in the *Montado*, we will use the social discourses as entry point that may be disentangled in the assertions expressed by the various key actors influencing the land use and landscape management. A discourse can be defined as an entity of signs that attribute meaning to particular objects, subjects, and statements [7]. When analyzing land use management, one may find it useful to explore the underlying discourses because they define the statements that are seen as legitimate about a given space or topic among a group of farmers. Therefore, discourse analysis can illicit the tacit normative foundation and the routines that are taken for granted on which the farming system and their underpinning management paradigms are based. This turns discourse analysis into an important analytical tool to unfold how these power relations play out in the case study setting.

Thus, why we opted for discourses as analytical tools to identify the management paradigms under which decisions undertaken by actors on the ground can be better understood.

In this analysis, we have explored the argumentation that emerges when farmers and land managers describe and explain their individual conduct. Our intention was not to judge particular regimes of knowledge and truth, but rather the ambition was to unfold how particular discourses provide a meaning to the management decisions that are made by various actors, and therefore to the management paradigms that underpin such decisions. We, therefore, looked for the existence of discourses, which provide justification for the different management paradigms currently underpinning farmers' decisions in the *Montado*. We did so by using an approach that could be easily replicated in other cases where multiple views and interests have gradually or simultaneously driven various management options for land use change.

4.2 Methods and sources

To achieve this, we explored different sources collated over the past 10 years, which jointly provided us with a quite comprehensive insight into the discourses and management paradigms related to this particular land use system, in the region: a review of published and unpublished results from research projects undertaken over the last 10 years, where different management options and drivers have been discussed and explored with land managers and other stakeholders [10, 11].

This was added on top of an analysis of media sources since 1990s [15]. This analysis consisted in the consultation of media sources referring to the *Montado* in Portugal, which was published between 1992 (coinciding with the MacSharry CAP reform) and May 2006. To perform this analysis, we examined 19 sources of information related to the production sector (webpages, newspapers, and magazines), using the NVIVO11 text codification software, participatory observation. This has been performed throughout our participation in various meetings where land owners express their opinion: thematic workshops, the preparation of the Green Book of the *Montado*s [10], the Advisory Group of the Ministry of Agriculture for the reform of the CAP post-2013, participatory discussion forum on the *Montados* organized once a month since April 2016 at the University of Évora, and 25 in-depth interviews with *Montado* land owners and managers, as well as other stakeholders involved in its management. Finally, the results and key findings of three focus groups with land managers, one workshop with a variety of stakeholders, and a survey with over 150 farmers in the Alentejo [15] were also considered.

The projects from which we collated information for this target were the following: EU grant no. 635,577, H2020-SFS-2014-2 (SUFISA: Sustainable finance for sustainable agriculture and fisheries") and the FCT grant FCT-PTDC/CS643 GEO/110944/2009 (ATILA: "Transition pathways: assessing innovation in farm management strategies"). Furthermore, results shown in this chapter were also drawn from a research project funded by National Funds through the FCT—Foundation for Science and Technology under the project UID/AGR/00115/2013. Since key relevant results from these projects are still unpublished, they cannot be properly cited in the text, and thus to obtain further details about them, readers should await for the publication of on-going papers, databases, or reports currently under way.

4.3 Results and findings

We identified three discourses (**Table 2**) using a grounded approach of the diverse ideas expressed by different stakeholders. Central statements were initially identified, grouped, and regrouped, leading to the generation of the different discourses that could be considered as "entry points" to the management paradigms currently shaping the fate and direction of the *Montado*. These statements were then examined in relation to (a) how language is used to "construct" the ideas or interpret information, (b) their variability, looking for inconsistencies of meaning in the constructions, via constant comparison and exploitation of the assumptions they reveal, and (c) the implications of the particular accounts and what is achieved in terms of discourse throughout each of these accounts [27]. The discourses were first conceived as ideal types, then progressively grounded and more comprehensively identified, through our interpretation of the different sources. **Table 2** shows how the three discourses can be characterized and differentiated.

Within each of the three discourses (**Table 2**) and related management paradigms, quite different versions of the *Montado* are constructed. Such versions are underpinned by different perception of the conditions for agricultural development

MONTADO Agro-silvopastoral systems	Heritage farming	Modern production farming	Land stewardship farming
Time scales	Nineteenth century and first half of the twentieth century	Started during 1960s and 1970s of the twentieth century, with a break during the revolutionary period (1974–1975), and again rising, following the integration of Portugal in the European Union	Started following the agrienvironmental stage of the CAP in 1992, but mainly after 2000
Central concepts	Rural culture Land estate Mixed agro-silvopastoral system	Modernization Rationalization Specialization Income-based decisions	Nature conservation Market Societal services Farming as nature keeper
Scope and perspective	Generational perspective	Shorter term perspective (financial gains)	Longer term perspective (intergenerational)
Societal perceptions on farming	Farming is the backbone of regional society, and farming maintains the state of social and cultural structures	Conservative and state/EU dependent	Double: both as nature keeper and nature destroyer
Key actors	Large land owners and their families	EU support schemes and labor wages	NGOs, Media, and EU agri-environmental schemes

Discourses are in this case considered as language-based analytical tools useful to unravel the diverse management paradigms under which decisions are made by farmers and land managers.

Table 2.
Main contemporary discourses in the Portuguese Montado and their key characteristics

in the region and prescribe different and often conflicting management practices, which align with the different management paradigms. In the heritage discourse, the *Montado* is constructed as a set of agriculture and forestry management practices and as a unique cultural landscape overly acknowledged as heritage. The modern production discourse poses the *Montado* as a production and management system prone to be modernized and intensified, with potential for high-income generation for the land owners' benefit. Finally, the land stewardship discourse conceives the *Montado* as a forestry-based nature conservation system, which is a guarantee of environmental balance in the region. Stemming from our own research through the aforementioned projects via which the overall information was collected, it is apparent how the current dominant discourse in the context of the Portuguese *Montado* is still a modern production farming, which has so far marginalized the other two discourses.

This is reflected both within the farming community and in designing public policies. This trend is associated with prevailing power relations in the farming community, with the role of the very large national farmers′ unions and with the positions held by key public institutions. Such dominance is strongly driving a minor role for the two other discourses, a fact that can be traced in the farmer's narratives.

The modern production farming discourse is linked with a series of farming management practices and options that are generally connected with monofunctional production models, thus neglecting the internal complexity of the system. Furthermore, these practices are directly dependent of external factors such as markets, marketing, and payment schemes. This represents a step backward in farm management skills and empowerment and in a loss of valuable-accumulated empirical knowledge, especially in relation to the farmers' capacity to deal with complexity, a capacity that can be considered essential to advance toward increased levels of resilience and sustainability [1].

As for the land stewardship discourse, numerous administrative and financial barriers exist toward operationalizing a number of management strategies that are essential under this paradigm, especially those related to nature and landscape conservation, making it difficult for farmers to embrace their role as land stewards. This is so despite that many farmers perceive the potential of their role as heritage stewards, a role with which they share long-term perspectives focusing on securing the well-being of future generations.

Still, concerns about the economic viability of farm units, which are strongly dominant in the modern production discourse, do also create barriers toward any novel management strategies.

Although environmental stewardship is clearly a rising discourse and management paradigm in the context of the *Montado* in Alentejo, it is still mostly prevalent in the national and international policy scales—and scarcely expressed in the policy tools acting at the farm level. Furthermore, it is mainly found with the regional to local policy institutions traditionally in conflict with agriculture, namely nature conservation, and also in actors with an integrated territorial focus, such as landscape and regional planning or tourism. Indeed, it is a discourse that is yet to be adopted by the farmers.

4.4 Key implications and reflections

National and international experts have long advocated in the media for a shift in the current productivist policies toward creating the conditions to maintain traditional farming systems, such as the *Montado*. However, it is commonly accepted by the farming sector this should not undermine the potential for Portugal to pursue increased production goals, which tightly fit with the modern production farming discourse. The question thus remains on how traditional systems such as the *Montado* can be protected while still contributing to increased production goals, with proposals such as sustainable intensification, land sparring, and others being at time made, but without much empirical evidence either on their implementation nor on their potential benefits. For advocates of the latter options, it will thus not be necessary to choose between drastically conflicting options, and there shall be a place for both intensive and extensive agricultural practices, as currently encountered in some areas of Portugal.

However, the intensive production systems that are advocated under the modern production discourse directly clash with the multifunctionality that is inherent to the *Montado* landscapes.

Resuming, it emerged clearly from our analysis that the three discourses, and related management paradigms, identified address three fundamental aspects of the Portuguese silvopastoral system: the historical and heritage value, the production role, and the environmental benefits.

Based on our analysis, we argue that each discourse and management paradigm in isolation fails to propose a uniquely valid solution to the sustainable reproduction of the system. A further reproduction of the *Montado* system is required that combines the three aforementioned dimensions. From an economic and organizational standpoint, this needs to draw on different support systems to develop a new discourse and paradigm that can encompass all these different aspects.

Furthermore, it is also clear that the co-existence in time and space of these different discourses, and the fuzziness in their boundaries, creates tensions in the farmers and land managers. In a way, these actors are often placed in between two or all of them but draw on several for justification of decisions in relation to different issues, mostly without recognizing the inherent contradictions in their management strategies.

They need to act simultaneously in different arenas, and different arenas favor different discourses. We found that too frequently, each individual *Montado* farmer finds himself divided among all three discourses or, more accurately, is placed in an internal conflict when undertaking management decisions. He may opt to be positioned within one dominant discourse and reproduce its values and contribute to a consistent narrative. And by doing this, even if he also relates to the other discourses, he identifies them as marginal, and will not be open to new combinations of the *Montado* components, nor for related management decisions.

Alas, the absence of a clear positioning as well as the conflicts between the separate discourses and management paradigms are constantly emerging at different levels of decision-making, ranging from the farm to the regional and national administrations, and also including negotiations with Brussels for the policy support mechanisms. With time, many public policy tools applied at the *Montado* have led to contradictory practices and decisions at the farm level. Consequently, besides the problem of the *Montado* quantitative and qualitative decay and related environmental degradation, an identity crisis is equally prevalent. This is driven by the fact that there a number of different visions of the future for this farming system that currently co-exist and that are mutually incompatible, potentially generating confusion among public and private actors acting at different levels of decision-making and responding to a number of different demands, both societal and economic. Examples of differences in such visions include the ambitions to expand or focus public and private investments on certain livestock species, particularly cattle, or to specialize in cork production, both of which are rotted in the modern production discourse and which are in sheer contrast, if not in plain conflict with the vision for attaining a truly multifunctional system, which alternatively underpins both the heritage and land stewardship discourses.

Resulting from these findings, It has become clearer that the multiple opportunities available for enhanced sustainability of complex land use systems such as the *Montado* may not be efficiently tackled should the currently prevalent modern production management paradigm persist. In response, in the following section, we discuss about future pathways for enhancing *Montado* sustainability options under alternative management paradigms.

5. Discussion: how to re-invent the sustainability of the *Montado*?

5.1 Why do we need to re-invent the sustainability of the *Montado*?

Resulting from our insights and research experiences in the Portuguese *Montado* over more than a decade, it is now emerging that the general trend toward qualitative and quantitative degradation will not be effectively addressed unless we obtain a clearer picture of the future direction of the system that can lead to enhanced sustainability. Furthermore, farmers and other key decision-makers will also need to be convinced that such picture is worth attaining, and the correct incentives and tools to achieve this need to be implemented. Alas, it is equally clear that the variety and imbalances of discourses and management paradigms that can be found co-existing hamper our potential to obtain and implement such a desired picture. Finally, a crucial argument needs to be made that sustainability in a system as complex as the *Montado* cannot be approached by addressing single issues on isolated topics, but that instead the entire system and the systemic interlinkages that produce unsustainable outcomes need to be jointly tackled. Thus, it seems clear that under a state-of-the-art scenario, the much-needed changes to improve sustainability will likely fail, and that we need to re-invent a future for the system that is based on novel approaches that consider changes in management paradigms and help advance toward enhancing both efficiency and sustainability.

5.2 Key changes required in management paradigms and related discourses for achieving improved sustainability

Regarding the first of these questions, no set of scenarios have been produced so far nationally for any components of the *Montado*, which is less so for the system as a whole. Nevertheless, official figures are available featuring recent and historic trends regarding issues critical for setting future scenarios, including tree cover, livestock composition and density, property structure, and human workforce [10]. Adding onto such scattered data, the analysis of discourses and management paradigms in this chapter provides an overall picture that may help in constructing scenarios.

According to our findings, tensions will likely persist between a set of management paradigms: the first and predominant of which modern production is influenced by financialization, free-trade-driven globalization, and economic competition, while an alternative one is now slowly emerging that reflects stewardship values, consideration of land and landscape as heritage, and value-adding through quality produce.

Indeed, the modern production discourse seems to be winning the battle both in the public (as reflected in the media) and political (as indicated through legislation being passed and plans approved) realms. This is consistent with the worldwide tensions between productivism and postproductivism [28] in agriculture, where despite the recent upsurging of the latter, the former is still clearly the dominant paradigm. This is actually the case even in the European context where postproductivism is now largely encouraged, both by policy-makers and by large components of the civil society [2]. The potential effects of the apparent resistance to move toward postproductivism will likely marginalize even further multifunctional farming systems such as the *Montado* that are in a clear competitive economic disadvantage in a context where economic efficiency is the ultimate goal of agriculture, but that could instead prove competitive under a postproductivist paradigm [1], where land stewardship could be considered as a valid management paradigm.

5.3 Getting there

In regard to the second of the questions posed (what needs to change?), we consider the improvement of coordination of scalar politics, as a crucial challenge that demands urgent improvements. We hereby consider scalar politics as re-defined by Lawhon and Patel in 2013 [29] in a context of globalisation and enhanced demand for sustainability. According to the pioneer arguments put together by Brenner in 1997 [30], unless better scalar coordination is achieved, the local initiatives that are now widely advocated for improved governance and sustainability of rural areas [31] will likely fail. This is a goal that entails incentives, both sociopolitical and economic, being improved.

To achieve better scalar coordination of governance and decision-making in *Montados*, one key aspect to tackle is to improve the currently policy and planning framework, which according to our findings is largely inefficient and lags an overall coherence and coordination. One could argue that such coherence could potentially be provided by an overarching plan that considers the complexity of *Montados* as a whole. However, according to the experience in the Spanish Dehesas, where a series of regional plans and regulations specifically targeting the Dehesas have already been approved and implemented for quite a while, it seems evident that this may be a necessary but a nonsufficient condition. While in the case of Extremadura, the pioneer region in testing the pathway of targeted policy integration, the main problems encountered relate to the lack of capacity to renovate an extremely outdated legal and policy framework [32], and in the case of Andalucía [33, 34], the key problems relate to the limitations encountered in the (largely strategic) planning instruments in place to connect the variety of scales across which decisions are made, and to engage the key stakeholders influencing land use change.

These are lessons from regions with a system similar to the *Montado* and with relatively equal socioecological conditions and challenges and thus should be indeed considered if the pathway of policy integration through planning for the Portuguese *Montado* was to be explored. On such regard, we argue that further efforts are required to improve the compatibility and mutual co-existence of the different alternatives currently available to decision-makers at different levels from the farm to the region. This should also be accompanied by specific plans that are operational and flexible enough to adapt to the specificities of the different contexts and typologies of *Montados*.

Furthermore, and even if these targets were to be specified in a plan, the heterogeneity of situations indicates that a well-crafted and carefully implemented participation scheme in the conception of the plan must firstly be prepared. This was actually how the Andalusian action plan from 2017 was crafted [34], defining some clear lessons and pathways forwarded for the Portuguese *Montado*. Drawing from such lessons, it seems now clearer that to be fully efficient, such a participatory scheme needs to be as inclusive as possible, engaging as many stakeholder groups as possible during many stages of the planning and policy-making process, especially those linking policies with management at the farm level, where the decisions are to be made.

However, these are all recommendations set at the wider policy level, and that therefore in principle, these operate in a different realm at which our analysis in this chapter has been conducted on which discourses and management paradigms realizing at the farm level are at the core of the discussion. However, unless we are capable of better linking together these policies and management decision levels, and focus on avoiding clashes between different management paradigms and underpinning discourses, securing sustainability for the Montado will become effectively impossible. In this sense, we consider the need to design a more efficient scalar governance system for the *Montado* as essential. Nevertheless, this is a "wicked challenge" [35], which will therefore only be effectively tackled through participatory and transdisciplinary schemes that transcend the strict agendas of individual research projects, farm plans, and personal research interests [36, 37]. Furthermore, any progress to be made in advancing and widening participatory decision-making will still need to be translated into effective policies and management schemes that can help move beyond the current leading paradigms and related discourses under which most farmers and land managers in the *Montado* currently operate.

As with the coordination of policies and planning, this may be a condition *sine quae non*, but it is also far from sufficient to achieve our goal to shift the various management problems and problems that we have encountered. Indeed, a number of barriers exist that may ultimately prevent the implementation of this recommendation from turning into a reality. These barriers include:

- the slow and highly bureaucratic inertia under which largely siloed policy making (e.g., agricultural vs. environmental policies) still operate in Portugal; the inability of such specialized policy frameworks to address problems faced in inherently multifunctional systems such as the *Montado;*

- a lack of the social and cultural conditions (e.g., associativism) required to foster cooperation among managers and farmers following different management paradigms, the reduced levels of market power [15] held by a community of farmers that are extremely dependent on public subsidies, and international markets in the short and medium terms;

- the difficulties for a long-term-driven system such as the *Montado* to compete under a modern production discourse and paradigm that emphasize short-term profits and competitiveness for land use;

- and the related difficulties in operationalizing sustainability objectives that may result in short-term loss of financial gains, and the lack of clarity and frequent misuse of controversial terms that may end up justifying actions and decisions with clear negative impacts for the system's sustainability.

This last point may be actually illustrated through the highly contested concept of "sustainable intensification" [38], a concept that largely related to that of soft sustainability, which is now considered as illegitimate in the framework of sustainable farming [39], but that has nevertheless been used as a cornerstone of the propaganda machine developed within the modern farming discourse to counteract the rising strength of land stewardship and other alternative management paradigms lately to arise.

In view of all of these challenges for the *Montado*, it is thus yet unclear whether a realistic scope exists to shift the current management paradigms and overcome the aforementioned barriers and move toward enhanced sustainability standards.

Although the diversity of situations indicated prevents us from indicating to "one-size-fits-all" type strategies, there a few general principles that may help drive change in the correct direction.

This includes expanding the current efforts in transdisciplinary knowledge co-construction toward enhancing linkages between management and the planning and policy-making processes, thus using academia as a real bridge between policies and management practices. This is a goal that could also benefit from improved extension services that are specialized in providing information for the *Montado* farmers. Ultimately, this may help alleviate tensions arising among defendants of the various management paradigms currently in place and help streamline a more coherent, clear, and sustainable future for the system that, however, considers and reflects the diversity of situations, mindsets, and aspirations of farmers and managers in the *Montado*.

Reflecting the aforementioned necessity to better reflect and respond to diverse conditions, it might be useful to consider setting standards that are better tailored to specific local conditions when setting management requirements as prerequisites for farmers to receiving public funding, as is currently the case with Pillar-II subsidies. Furthermore, current proposals and experiences in CAP Pillar II to move beyond practice-based payments and onto result-based payments could help trigger the much-needed paradigm questioning and shift that is hereby advocated to improve sustainability.

In this sense, extension services, in close coordination with other relevant public authorities such as those dealing with nature and landscape designations and environmental standards (ICNF), could also function as responsible institutions to make sure farmers and land managers are adequately informed and engaged about their many possibilities and limitations, and also encouraged and empowered to help trigger any paradigmatic changes needed.

A last point worthy of discussion would relate to the opportunities lately opening to marketize and underline the diverse cultural and environmental benefits of the *Montado* system by emphasizing product quality, thereby giving farmers an incentive to a more sustainable production, thereby decoupling farmers from the world market. This is done in many other places of Europe, via genetic (PDO) and geographical (PGI) varietal protection schemes, and is very much aligned with the land stewardship discourse, which is in this way assigned to a marketing and value-adding tool that may help render it more attractive and competitive in the view of farmers and land managers, ultimately helping advance beyond the productivist approach that is currently dominant.

6. Conclusions

We began this chapter by outlining the complex, or even wicked, challenges that are associated with a silvopastoral system, the Portuguese *Montado*. This is a system that is characterized by its great potential for improving sustainability standards by helping reconcile farming productivity with biodiversity conservation, landscape-cultural heritage protection, and local economic development. Despite this, the system is currently suffering from a strong qualitative and quantitative degradation, and neither efficient policies nor management strategies are in place that is able to embrace the complexity of the system. We have argued that this is due to conflicts and imbalances and trade-offs between various management paradigms, which need to be better unraveled and understood.

To achieve such goal, we applied a discourse analysis, which allowed us to identify and characterize three distinctive management paradigms and corresponding discourses that have co-existed, generating tensions in the past 100 years. Currently, it seems clear that a modern production paradigm and discourse is still prevalent, which is partly at the root of the unsustainable trajectories that the *Montado* is mostly following. In response, a land stewardship alternative paradigm is slowly rising that provides with some expectations as to more sustainable futures for the system, while remnants of a heritage paradigm still subsist in association with the conservative societal views that characterize much of the Portuguese countryside. Tensions arising among such management paradigms and related discourses are very much aligned with the transition from a productivist toward a postproductivist countryside and agriculture.

Overall, the paradigmatic shift in management that is required to secure the future sustainability of the Portuguese *Montado* is not happening. Neither policies, nor farming mind-sets nor financial instruments seem to be adequately placed to help reverse current trends of decline and degradation in the system. Nevertheless, some tips and directions for the future could be identified that may help achieve improvements and that are related to improving governance and scalar coordination, re-defining financial and policy incentives so that they are reflective of the diversity of situations potentially encountered, and better supporting, informing and guiding farmers and land managers operating on the ground through reformed extension services and knowledge-co-construction strategies that can help them consider a wider range of factors and opportunities. Whether these will ultimately be operational on the ground still remains an enigma, but at least, the baseline is now enriched in a way that, even if it does not ultimately lead to more sustainable pathways, may provide decision-makers a better picture of what the possible pathways and options may be.

Acknowledgements

Research shown in this chapter has received funds from the EU's Horizon 2020 research and innovation program under grant agreement no: 635577 (www.sufisa.eu).

Author details

José Muñoz-Rojas[1]*, Teresa Pinto-Correia[1], Martin Hvarregaard Thorsoe[2] and Egon Noe[3]

1 ICAAM—Institute of Mediterranean Agricultural and Environmental Sciences, University of Évora, Pólo da Mitra, Évora, Portugal

2 Department of Agroecology, Aarhus University, Agroecology, AU Foulum, Tjele, Denmark

3 Department of Sociology, Environmental and Business Economics, Danish Centre for Rural Research, University of South Denmark, Esbjerg, Denmark

*Address all correspondence to: jmrojas@uevora.pt

References

[1] Wilson GA. Multifunctional Agriculture: A Transition Theory Perspective. New York; Wallingford: Oxford University Press; CAB International; 2007. 384 pp. ISBN: 978-1845932565

[2] Sutherland LA, Darnhofer I, Wilson G, Zagata L. Transition Pathways towards Sustainability in Agriculture. Wallingford: CAB International; 2015. 246 pp. ISBN: 9781780642192

[3] Slee B, Pinto-Correia T. Understanding the diversity of European rural areas. In: Sutherland LA, Darnhofer I, Wilson G, Zagata L, editors. Transition Pathways towards Sustainability in Agriculture. Wallingford: CAB International; 2015. pp. 33-50. ISBN: 9781780642192

[4] Erjavec K, Erjavec E. Changing EU agricultural policy discourses? The discourse analysis of Commissioner's speeches 2000-2007. Food Policy. 2009;**34**:218-226. DOI: 10.1016/j.foodpol.2008.10.009

[5] Darnhofer I. Socio-technical transitions in farming: Key concepts. In: Sutherland LA, Darnhofer I, Wilson G, Zagata L, editors. Transition Pathways towards Sustainability in Agriculture. Wallingford: CAB International; 2015. pp. 17-32. ISBN: 13: 978-1780642192

[6] Godinho S, Guiomar N, Machado R, et al. Assessment of environment, land management, and spatial variables on recent changes in *Montado* land cover in southern Portugal. Agroforestry Systems. 2016;**90**:177. DOI: 10.1007/s10457-014-9757-7

[7] Foucault M. The subject and power. Critical Inquiry. 1982;**8**(4):777-795. Available from: https://www.jstor.org/stable/1343197?seq=1#metadata_info_tab_contents

[8] Bugalho M, Pinto-Correia T, Pulido F. Human use of natural capital generates cultural and other ecosystem services in Montado and dehesa oak woodlands. In: Paracchini ML, Zingari PC, Blasi C, editors. Re-connecting Natural and Cultural Capital. Contributions from Science and Policy. Luxembourg: Publications Office of the European Union; 2018. ISBN: 978-92-79-59948-4

[9] Moreno G, Franca A, Godinho S, Pinto-Correia T. Multifunctionality and dynamics of silvo-pastoral systems. Options Méditerranéennes, A. 2014;**109**:421-436. In: Forage resources and ecosystem services provided by Mountain and Mediterranean grasslands and rangelands. Available from: http://agris.fao.org/agris-search/search.do?recordID=QC2017600093

[10] Pinto-Correia MT, Ribeiro N, Potes J, editors. Livro Verde dos Montados. ICAAM-Universidade de Évora; 2013. 61 pp. Available from: https://dspace.uevora.pt/rdpc/bitstream/10174/10116/1/Livro%20Verde%20dos%20*Montado*s_Versao%20online%20%202013.pdf

[11] Ferraz-de-Oliveira I, Azeda C, Pinto-Correia T. Management of *Montados* and Dehesas for high nature value: An interdisciplinary pathway. Agroforestry Systems. 2016;**90**:1-6. DOI: 10.1007/s10457-016-9900-8

[12] Pinto-Correia T, Azeda C. Public policies creating tensions in *Montado* management models: Insights from farmers' representations. Land Use Policy. 2017;**64**:76-82. DOI: 10.1016/j.landusepol.2017.02.029

[13] Pinto-Correia T, Godinho S. Changing agriculture—Changing landscape: What is going on in the high valued *Montado* landscapes of Southern Portugal? In: Ortiz-Miranda D, Moragues-Faus AM, Arnalte-Alegre E, editors. Agriculture

in Mediterranean Europe Between old and new paradigms. Research in Rural Sociology and Development. Vol. 19. Bingley, UK: Emerald Insight; 2013. pp. 75-90. ISBN: 978-1-78190-597-5

[14] Instituto Nacional de Conservação da Natureza e Florestas (ICNF). Inventario Nacional Florestal. Áreas dos usos do solo e das espécies florestais de Portugal continental. Resultados preliminares. 2013. 34 pp. Available from: http://www2.icnf.pt/portal/florestas/ifn/resource/ficheiros/ifn/ifn6-res-prelimv1-1

[15] Muñoz-Rojas J, Pinto-Correia T. SUFISA National Report-Portugal (Deliverable 2.2). 2018. Available from: https://www.sufisa.eu/wp-content/uploads/2018/09/D_2.2-Portugal-National-Report.pdf

[16] Biely K, Maes D, Van Passel S. Market power extended: From foucault to meadows. Sustainability. 2018;**10**(8):2843. DOI: 10.3390/su10082843

[17] Instituto Nacional de Estadística. Dados dos censos de população por NUTS II. 2018. Available from: https://www.ine.pt/xportal/xmain?xpid=INE&xpgid=ine_base_dados

[18] CCDR—Alentejo Regional Coordination and Development Commission. Regional Strategy for Smart Specialization in Alentejo. 2014. Available from: https://ec.europa.eu/growth/tools-databases/regional-innovation-monitor/policy-document/alentejo-2020-%E2%80%93-regional-operational-programme

[19] CCDR—Alentejo Regional Coordination and Development Commission. Alentejo 2020—Regional Operational Programme. 2014. Available from: https://ec.europa.eu/growth/tools-databases/regional-innovation-monitor/policy-document/alentejo-2020-%E2%80%93-regional-operational-programme

[20] Governo de Portugal. Resolução do Conselho de Ministros-Plan Regional de Ordenamento Territorial do Alentejo-n.° 53/2010 Diário da República, 1.ª série—N.° 148—2 de Agosto de 2010. 2010. Available from: http://webb.ccdr-a.gov.pt/docs/ordenamento/RCM_53-2010_2Ago.pdf

[21] Governo de Portugal. Decreto-Lei n.° 155/2004, de 30 de Junho. D.R. n.° 152, Série I-A. Altera o Decreto-Lei n.° 169/2001, de 25 de Maio, que estabelece as Medidas de Protecção ao Sobreiro e à Azinheira - altera os arts. 1.°, 3.° e 13.° do DL n.° 169/2001, de 25 de Maio e adita-lhe o art. 1.°-A. 2004. Available from: https://dre.pt/application/dir/pdf1sdip/2004/06/152A00/39673968.pdf]

[22] Governo de Portugal. Decreto-lei 42/2007, de 22 de Fevereiro. Define o regime jurídico aplicável à gestão, exploração, manutenção e conservação das infra-estruturas que integram o empreendimento de fins múltiplos de Alqueva EFMA. 2007. Available from: https://dre.tretas.org/dre/206946/decreto-lei-42-2007-de-22-de-fevereiro

[23] Gallent N, Scott M. Rural Planning and Development. London: Routledge; 2017. 202 pp. ISBN: 9781138016347 - CAT# Y165113

[24] Governo de Portugal. Procede à primeira alteração ao Decreto-Lei n.° 73/2009, de 31 de março, que aprova o regime jurídico da Reserva Agrícola Nacional. DECRETO-LEI N.° 199/2015 - DIÁRIO DA REPÚBLICA N.° 181/2015, SÉRIE I DE 2015-09-16. 2015. Available from: ran.drapc.min-agricultura.pt/docs/decreto_lei_199_2015.pdf

[25] Municipio de Montemor-o-Novo. Aviso 3453/2011, de 1 de Fevereiro. Plano de Intervenção no Espaço Rural do Sítio de Monfurado—PIERSM. 2011. Available from: https://dre.tretas.org/dre/1222434/aviso-3453-2011-de-1-de-fevereiro

[26] Berke PR, Conroy MM. Are we planning for sustainable development? An evaluation of 30 comprehensive plans. Journal of the American Planning Association. 2000;**66**:21-33. DOI: 10.1080/01944360008976081

[27] Silverman D. Interpreting Qualitative Data. 4th ed. SAGE Publishers; 2011. 520 pp. ISBN-13: 978-0857024213

[28] Wilson GA. From productivism to post-productivism... and back again? Exploring the (un)changed natural and mental landscapes of European agriculture. Transactions of the Institute of British Geographers. 2001;**26**(1):77-102. DOI: doi.org/10.1111/1475-5661.00007

[29] Lawhon M, Patel Z. Scalar politics and local sustainability: rethinking governance and justice in an era of political and environmental change. Environment and Planning C: Politics and Space. 2013;**31**(6):1048-1062. DOI: 10.1068/c12273

[30] Brenner N. Global, fragmented, hierarchical: Henri Lefebvre's geographies of globalization. Public Culture. 1997;**24**:135-167. DOI: 10.1215/08992363-10-1-135

[31] Cowell R. Localism and the environment: effective transition for a sustainability transition?, In: Davoudi S, Mandanipour A, editors. Re-considering Localism. London (UK): RTPI; 2015. pp. 216-237. ISBN: 978-0415735612

[32] Junta de Extremadura. Ley 1/1986, de 2 de mayo, sobre la Dehesa en Extremadura. 1986. Available from: https://www.boe.es/buscar/pdf/1986/BOE-A-1986-19748-consolidado.pdf

[33] Junta de Andalucía. Ley 7/2010, de 14 de julio, para la Dehesa. 2010. Available from: https://www.juntadeandalucia.es/boja/2010/144/1

[34] Junta de Andalucía. Decreto 172/2017, de 24 de octubre, por el que se aprueba el Plan Director de las Dehesas de Andalucía, se crea su Comité de Seguimiento y se modifica el Decreto 57/2011, de 15 de marzo, por el que se regula la Comisión Andaluza para la Dehesa y el Decreto 530/2004, de 16 de noviembre, por el que se regula la composición, las funciones y el régimen de funcionamiento del Consejo Andaluz de Biodiversidad. 2017. Available from: https://www.juntadeandalucia.es/boja/2017/207/BOJA17-207-00007-18430-01_00123501.pdf

[35] Duckett D, Feliciano D, Martin-Ortega J, Munoz-Rojas J. Tackling wicked environmental problems: The discourse and its influence on praxis in Scotland. Landscape and Urban Planning. 2016;**154**:44-56. https://doi.org/10.1016/j.landurbplan.2016.03.015

[36] Guimarães MH, Fonseca C, Gonzalez C, Pinto-Correia T. Reflecting on collaborative research into the sustainability of mediterranean agriculture: A case study using a systematization of experiences approach. Journal of Research Practice. 2017;**13**(1):Article M1

[37] Guimarães MH, Guiomar N, Surová D, Godinho S, Pinto Correia T, Sandberg A, et al. Structuring wicked problems in transdisciplinary research using the social–ecological systems framework: An application to the Montado system, Alentejo, Portugal. Journal of Cleaner Production. 2018;**191**:417-428. DOI: 10.1016/j.jclepro.2018.04.200

[38] Petersen B, Snapp S. What is sustainable intensification? Views from experts. Land Use Policy. 2015;**46**:1-10. DOI: 10.1016/j.landusepol.2015.02.002

[39] Biely K, Maes D, Van Passel S. The idea of weak sustainability is illegitimate. Environment, Development and Sustainability. 2018, 2018;**20**:223. DOI: 10.1007/s10668-016-9878-4